Development of a low-cost alternative for metal removal from textile wastewater

T0186372

Thesis committee

Thesis supervisor

Prof. dr. ir. P.N.L. Lens
Professor of Biotechnology
UNESCO-IHE Institute for Water Education
Delft, the Netherlands

Thesis co-supervisor

dr. ir. D.P.L. Rousseau
Senior Lecturer Environmental Engineering
UNESCO-IHE Institute for Water Education
Delft, the Netherlands

Other members

Prof. dr. ir. C.J.N. Buisman
Wageningen University
Wageningen, the Netherlands

Prof. dr. ir. G. Du Laing
Ghent University
Ghent, Belgium

Prof. dr. ir. E. van Hullebusch
Université Paris-Est
Paris, France

dr. S.K. Sharma
UNESCO-IHE Institute for Water Education
Delft, the Netherlands

Development of a low-cost alternative for metal removal from textile wastewater

DISSERTATION
Submitted in fulfillment of the requirement of
the Academic Board of Wageningen University and
the Academic Board of the UNESCO-IHE Institute for Water Education
for the degree of DOCTOR
to be defended in public
on Friday, 29 June 2012 at 16:30 hrs
in Delft, the Netherlands

by

Christian SEKOMO BIRAME
born in Lubumbashi, Democratic Republic of Congo (D.R.C)

CRC Press/Balkema is an imprint of the Taylor & Francis Group, an informa business

© 2012, Christian Sekomo Birame

All rights reserved. No part of this publication or the information contained herein may be reproduced, stored in a retrieval system, or transmitted in any form or by any means, electronic, mechanical, by photocopying, recording or otherwise, without written prior permission from the publishers.

Although care is taken to ensure the integrity and quality of this publication and the information therein, no responsibility is assumed by the publishers nor the author for any damage to property or persons as a result of operation or use of this publication and/or the information contained herein.

Published by:
CRC Press/Balkema
PO Box 447, 2300 AK Leiden, the Netherlands
e-mail: Pub.NL@taylorandfrancis.com
www.crcpress.com - www.taylorandfrancis.co.uk - www.ba.balkema.nl

ISBN 978-0-415-64158-6 (Taylor & Francis Group)
ISBN 978-94-6173-296-5 (Wageningen University)

Contents

Dedication

To my father who died early in 1981 without seeing this great achievement,

To my grandmother, grandfather, aunts, uncles, brother and cousins who prematurely died during the 1994 genocide in Rwanda,

May our almighty God grant your souls to rest in peace forever.

Acknowledgements

I would like to express my gratitude to my promoter Professor Dr. ir. P.N.L. Lens together with my mentor Senior lecturer Dr. ir. D.P.L. Rousseau of UNESCO-IHE Institute for Water Education, for their guidance, criticisms and their valued advices are highly appreciated. My special thanks to Professor H.J. Gijzen who was my first promoter at the beginning of my PhD studies, your bright ideas have been a great help and inspiration. Many thanks also to Professor Dr. ir. Innocent Nhapi and Dr. ir. Wali Umaru Garba coordinators of the Water Resources and Environmental Management (WREM) program at NUR for their administrative and supporting roles. I would like also to extend my gratitude to all UNESCO-IHE laboratory staff for their support during my experimentation period in the laboratory. The part of the work conducted by MSc students (Nkuranga, Hutriadi, Saleh, Gakwavu, Mwabi and Kagisha) is gratefully acknowledged. Thanks also to my fellow PhD colleagues from UNESCO-IHE, particularly those of the National University of Rwanda (Uwamariya Valentine, Munyaneza Omar, Uwimana Abias and Ingabire Dominique), pursuing their studies at UNESCO-IHE. It was a valuable experience to work with enthusiastic and dedicated colleagues and staff.

I gratefully acknowledge the financial support of NUFFIC, the Dutch government, the WREM Program and the National University of Rwanda in general for the opportunity offered for further studies at PhD level. Last, but certainly not least, my deeply felt sincere thanks go to my wife Aline Kabano, my daughter Ariane Birame Uwase, my aunt Louise Sanitas Sekomo and other family members for their patience, endurance and encouragement that have been useful throughout my PhD studies.

Abstract

Economic development, urbanization and population growth are three parameters that are influencing the quality of water nowadays. Textile wastewater is a typical industrial wastewater generated by human activities. This alkaline wastewater contains many pollutants and has high BOD and COD loads. Pollutants in textile effluents include suspended solids, mineral oils (e.g. antifoaming agents, grease, spinning lubricants, non-biodegradable or low biodegradable surfactants) and other organic compounds, including phenols from wet finishing processes (e.g. dyeing), and halogenated organics from solvent use in bleaching. Effluent from dyeing processes is colored and may contain significant concentrations of heavy metals (e.g. chromium, copper, zinc, lead, or nickel). The present study is focusing on heavy metal pollution found in textile wastewater because of their toxic effect in the environment and most studies on textile wastewater focus on the removal of organic pollutants.

In Rwanda, textile wastewater contributes to heavy metal pollution in the swamps receiving that water. Chapter 3 presents the assessment of metal pollution in a swamp receiving textile wastewater from UTEXRWA in Kigali (Rwanda). The Nyabugogo swamp is a natural wetland located in Kigali City (Rwanda). This wetland receives all kinds of untreated wastewaters from the city. The assessment of heavy metal pollution (Cd, Cr, Cu, Pb and Zn) included all environmental compartments of the swamp: water and sediment, the dominant plant species *Cyperus papyrus*, fish (*Clarias* sp. and *Oreochromis* sp.) and Oligochaetes. Cr, Cu and Zn concentrations in the water were generally below the WHO (2008) drinking water quality guidelines, whereas Cd and Pb were consistently above these limits. Except Cd, all metal concentrations were below the threshold levels for irrigation. The highest metal accumulation occurred in the sediment with up to 4.2 mg/kg for Cd, 68 mg/kg for Cu, 58.3 mg/kg for Pb and 188.0 mg/kg for Zn, followed by accumulation in the roots of *Cyperus papyrus* with up to 4.2 mg/kg for Cd, 45.8 mg/kg for Cr, 29.7 mg/kg for Cu and 56.1 mg/kg for Pb. Except Cu and Zn, other heavy metal (Cd, Cr and Pb) concentrations were high in *Clarias* sp., the *Oreochromis* sp. and the Oligochaetes. Therefore, there is a human health concern for people using water and products from the swamp.

From the results of chapter 3, there is a need for the development of a cheap, reliable technology applicable in developing countries for heavy metal removal from polluted water bodies. This research aimed at improving the heavy metal removal from textile wastewater using an integrated system for wastewater treatment. The integrated system consists in a combination of an anaerobic reactor as the main treatment step, followed by a polishing step composed by a macrophyte stabilization pond. The main treatment step will remove a large quantity of heavy metals, mainly by adsorption combined to metal sulfide precipitation. The polishing step will be achieved by phytoremediation where removal will be conducted by algae and plants (duckweed and water hyacinth). The integrated system will show the complementary action of both systems in metal removal.

In chapter 4, based on the results of the assessment of heavy metal concentrations in the swamp, we decided to test other plant species, i.e. duckweed, and to compare it with algae as abiotic conditions will not be the same in these systems. Two treatment systems (algae and duckweed ponds) were tested based on the hypothesis that different physico-chemical conditions (pH, redox potential and dissolved oxygen) would lead to differences in metal removal. These ponds have been operated at a hydraulic retention time of 7 days and under two different metal loading rates and light regimes (16/8 hours light/darkness and 24 hours light). Results revealed that Cr removal rates were 94 % for the duckweed ponds and 98 % for the algal ponds, indifferently of the metal loading rate and light regime. No effect of pond type could be demonstrated for Zn removal. Under the 16/8 light regime, Zn removal proceeded well (~ 70 %) at a low metal loading rate, but dropped to below 40 % at the higher metal loading rate. The removal efficiency rose back to 80 % at the higher metal loading rate but under 24 hours light regime. Pb, Cd and Cu all showed relatively similar patterns with removal efficiencies of 36% and 33% for Pb, 33% and 21% for Cd and 27% and 29% for Cu in the duckweed and algal ponds, respectively. This indicates that both treatment systems are not very suitable as a polishing step for removing these heavy metals. Despite the significant differences in terms of physico-chemical conditions, differences in metal removal efficiency between algal and duckweed ponds were rather small.

In chapter 5, from results of the previous chapter, we opted to look for alternative materials for metal removal from wastewater. Therefore, a new adsorbent material the Gisenyi volcanic rock found in Northern Rwanda was tested as potential adsorbent for metallic ions (Cd, Cu, Pb and Zn) from wastewater. Adsorption has been conducted under a variety of experimental conditions (initial metal concentration varied from 1 – 50 mg/L, adsorbent dosage 4 g/L, solid / liquid ratio of 1:250, contact time 120 hours, particle size 250 – 900 μm). The adsorption process was optimal at near-neutral pH 6. The maximal adsorption capacity was 6.23 mg/g, 10.87 mg/g, 9.52 mg/g and 4.46 mg/g for Cd, Cu, Pb and Zn, respectively. The adsorption process proceeded via a fast initial metal uptake during the first 6 hours, followed by slow uptake and equilibrium after 24 hours. Data fitted well the pseudo second order kinetic model. Equilibrium experiments showed that the adsorbent has a high affinity for Cu and Pb, followed by Cd and Zn. Furthermore, the rock is a stable sorbent that can be reused in multiple sorption–desorption–regeneration cycles. Therefore, the Gisenyi volcanic rock was found to be a promising adsorbent for heavy metal removal from industrial wastewater contaminated with heavy metals.

Results from chapter 5 showed the potential of the Gisenyi volcanic rock as sorbent. However, to overcome the saturation problem of the bonding sites of the sorbent, we developed and tested an integrated system for metal removal in chapter 6. That integrated system for heavy metal removal was a combination of an Upflow Anaerobic Packed Bed reactor (UAPBR) filled with the Gisenyi volcanic rock as adsorbent and attachment surface for bacterial growth constituting the main heavy metal removal step. This step was further combined to a water hyacinth pond as polishing step. The integrated set-up was operated at the NUR experimental facility (temperature = 25 °C, monthly solar radiation average varying between 4.3 to 5.2 kWh / m^2). Two different feeding regimes were applied using low (5 mg/L of heavy metal each) and high (10 mg/L of heavy metal each) metal concentrations in the wastewater. After a start up and acclimatization period of 44 days, each regime was operated for a period of 10 days with a hydraulic retention time of 1 day. Good removal efficiencies of at least 86 % have been achieved for both low and high strength wastewater. The bioreactor performance was not much affected when the columns were operated under high strength heavy metal concentrations. A subsequent water hyacinth pond with a hydraulic retention time of 1 day removed an additional 61 % Cd, 59 % Cu, 49 % Pb and 42 % Zn, showing its importance as polishing step. The water hyacinth plant in the post treatment step accumulated heavy metals mainly in the root system. Overall metal removal efficiencies at the outlet of the integrated system were 98 % for Cd, 99 % for Cu, 98 % for Pb and 84 % for Zn. Therefore, the integrated system can be used as an alternative treatment system for metal-polluted wastewater for developing countries.

The development of a low cost alternative for metal removal from wastewater showed how efficient such a treatment system can be. Metal sulfide precipitation as long term removal mechanism showed that removal efficiencies exceeding 90 % can be reached in the bioreactor. The use of algal and duckweed ponds as alternative for water hyacinth plants showed no differences based on abiotic differences. The performance of both algal and duckweed systems was close and these systems are well suited as polishing step for wastewaters containing low metal concentrations. The integrated system for heavy metal removal showed how two complementary systems for heavy metal removal can work in combination and a good overall removal performance can be achieved.

Chapter 1: General introduction

1.1. Water and sanitation in Rwanda

Rwanda is a country located in the Great Lakes Region of Africa. Its topography gradually rises from the East at an average altitude of 1,250 m to the North and West where it culminates in a mountain range called "Congo-Nile Ridge" varying from 2.200 m to 3,000 m and a volcano formation, the highest volcano being 4,507 m high. This topography is characterized by a vast number of hills and mountains, a fact which results in high soil erosion and loss of water. Rwanda possesses a dense hydrographical network. Lakes occupy 128,190 ha; rivers cover an area of 7,260 ha and water in wetlands and valleys cover a total of 77,000 ha. The country is divided by a water divide line called Congo-Nile Ridge. To the West of this line lies the Congo River Basin which covers 33% of the national territory, which receives 10% of the total national waters. To the East lies the Nile River Basin, whose area covering 67% of the Rwandan territory and delivers 90% of the national waters [Ministry of Lands, Environment, Forests, Water and Mines (MINITERE, 2004b)].

Rwanda's population in 2008 was estimated at 9,831,501 inhabitants with an annual growth rate of about 3.8 %. The population is therefore expected to rise to about 13 million inhabitants by 2020. Given its small land surface area of approximately 26,338 square kilometres, Rwanda is one of the most densely populated countries in Africa, with a population density estimated at 373 inhabitants per square kilometre [Ministry of Natural Resources (MINIRENA, 2011)]. Sanitation has traditionally lagged behind developments in water supply yet these two areas are crucial to guarantee public health. In Rwanda, 55% of the rural population and 69 % of urban dwellers have access to safe drinking water while statistics also show that a meagre 10% of Kigali city dwellers have access to sanitation services, while only 8% of rural population has adequate sanitation facilities [Ministry of Finance and Planning (MINECOFIN, 2003)]. Kigali City, the capital of Rwanda is growing rapidly at about 5% per annum, dramatically affecting the city landscape, with infrastructure now failing to cope with the loads imposed upon them. The sanitation facilities in the city are a serious cause of concern with more than 60% of the population using onsite facilities, mainly pit latrines. Of these, about 17% do not have pit latrines of their own, but share with neighbours. About 12% of Kigali's inhabitants use conventional septic tanks, whilst less than 1% is connected to five small wastewater treatment plants. Actually, a feasibility study for constructing a centralized sewer network and an activated sludge system for wastewater treatment in Kigali city has been completed. In a near future, the implementation phase is expected to be launched.

The country recognizes the importance of water and sanitation for the improvement of the living conditions of its population. Thus, better water resources management inevitably contributes to the reduction of poverty and to socio-economic development of the country. Since 1996, tremendous efforts have been made to launch a national policy for the management of the water and sanitation sector that consists of strategies and programmes for the construction and rehabilitation of human resources, social and economic infrastructures, and, to finally develop a long term vision in order to provide better guidance for the development and the coordination in this sector (MINITERE, 2004b).

Based on the sector-based policy project on water and sanitation developed in 1992, revised in 1997 and as well in 2001, Rwanda consequently formulated a new policy which defines guidelines for efficient use of resources which also integrates new aspects such as decentralisation, participatory approach, privatisation and funding through a programatic approach. This policy is in harmony with the MDG objectives and the 2020 Vision of the Rwanda Government which cater that all of its population will have access to drinkable water and to sanitation services by the year 2020 (MINITERE, 2004b). Furthermore, 90 % of Rwandan population depends on agriculture, with the poorest depending for livelihoods on forests, fishing and wetlands. About 30 % of morbidity in Rwanda is due to environmental causes and 20 % of child mortality is due to diarrhea, cholera and related diseases causes by polluted water and lack of sanitation [Ministry of Agriculture and Animal Resources (MINAGRI, 2007)].

The Government of Rwanda had put in place many regulations and the most important is the law N° 04 / 2005 of 08 / 04 / 2005 determining the modalities of protection, conservation and promotion of the environment in Rwanda (MINITERE, 2004a). That law determines the modalities of protecting, conserving and promoting the environment in Rwanda. The law aims at:
  (i) conserving the environment, people and their habitats;
  (ii) setting up fundamental principles related to protection of the environment, any means that may degrade the environment with the intention of promoting the natural resources, to discourage any hazardous and destructive means;
  (iii)promoting social welfare of the population considering equal distribution of the existing wealth;
  (iv)considering the durability of the resources with an emphasis especially on equal rights on present and future generations;
  (v) guarantee to all Rwandans sustainable development which does not harm the environment and the social welfare of the population;
  (vi)setting up strategies of protecting and reducing negative effects on the environment and replacing the degraded environment.

1.2. Textile wastewater and quality problems

Human industrial activities produce in general diverse pollutants. Depending on the industrial activities, these pollutants may be toxic organic, suspended solids and heavy metals (Eckenfelder, 1989; Al-Degs et al. 2000; Moreira et al., 2001; Chagas and Durrant, 2001; Toh et al., 2003). Depending on the processing, these wastewaters contain typical pollutants such as BOD, COD, oil and grease, TSS, nitrogen, phosphorus and heavy metals (Kadlec and Wallace, 2008). The pollution caused by the textile industry causes a particular threat to the Rwandan environment. The textile industry is playing a major role in the socio-economic life of Rwanda with job creation and payment of taxes to the government. However, it is important to mention that the same textile industry is also polluting the environment by dumping its wastewaters not well treated. Textile wastewater is related to wet operations, which are conducted during different parts of the textile manufacturing process (scouring, finishing operation, desizing, bleaching, mercerizing, dyeing, printing and mothproofing) (Mohan et al., 2005). These processes require the input of a wide range of chemicals and dyestuffs, which generally are organic compounds of complex structure (IFC & world Bank group, 2007).

Effluent streams from dyeing processes are typically hot and colored and may contain significant concentrations of heavy metals (chromium, copper, zinc, lead, or nickel) as reported by many researchers (Tsezos, 2001; Cheng et al., 2002; Merzouk et al., 2009; Fenglian and Qi, 2011). Industrial wastewater from natural fiber processing may contain pesticides used in prefinishing processes (cotton growing and animal fiber production), potential microbiological pollutants (bacteria, fungi, and other pathogens), and other contaminants (sheep marking dye, tar). This is particularly significant for animal fiber processing (IFC & world Bank group, 2007).

The quality of textile wastewater depends on the employed chemicals, coloring matters, and dyestuffs as well as the process itself. Depending on the season and the fashion, the composition of textile wastewater even of the same process changes often. The used chemicals and dyes lead to an inorganic as well as organic pollution of the wastewater (Brik et al., 2006). The main environmental concern is therefore about the amount of water discharged and the chemical load it carries. For illustration, for each ton of produced fabric $20 - 350 \ m^3$ of water are consumed, the rather wide range reflecting the variety of involved processes and process sequences (Brik et al., 2006). In order to reduce the environmental impact, discharge limits imposed on textile mills are becoming ever more stringent (Okeo-tex, 2006a, b, c). Moreover, in the future reuse of purified effluents will be of increasing relevance due to raising water prices as well as to preserve natural water resources. In the present research, our interest will focus on heavy metal pollution.

1.3. Need for research on cost effective textile wastewater treatment systems for developing countries

Rwanda is a developing country where wastewater treatment systems are not really operational even at small scale. Furthermore, the wastewater generated in Kigali city is dumped untreated into natural wetlands. This is causing a serious problem of water pollution, because all the wastewater is directly discharged into natural water bodies, without any prior treatment. Therefore, there is a need for monitoring the water quality within and surrounding the wetland receiving this polluted water, in order to assess the extent of the metal pollution. Furthermore, Rwanda as many developing countries is facing problems related with the treatment of domestic and industrial wastewaters, mainly due to high investment required for the installation and running cost of conventional treatment plants at the one hand and the need of skilled personnel that its also required at the other hand.

In trying to address this issue, this study proposed the collection of data in a natural wetland receiving wastewater coming from the only one textile industry UTEXRWA operating in the city of Kigali (Rwanda) with a focus on heavy metals pollution, specifically Cd, Cr, Cu, Pb and Zn. These metals found in wastewater are known to be toxic to human. They are discharged into the Nyabugogo wetland and are endangering the aquatic life in the Nyabugogo River (Figure 1–1). This textile industry consumes $7,000 \ m^3$ of clean water each month. After cloths processing, an average of $200 \ m^3/$ day of wastewater are discharged into the river flowing into the Nyabugogo swamp (Figure 1-1). The average chemical composition of this wastewater shows a concentration in heavy metals about 0.04; 0.15; 3.45; 0.35 and 1.26 mg/L respectively for Cd, Cu, Cr, Pb, and Zn. 2496 mg/L of COD; 196.5 mg/L of BOD; 43.4 mg/L of

NO_3^{-1}; 3.82 mg/L PO_4^{-3}; 435 mg/L of SO_4^{-2}; pH of 12.3 and a dark color in general. This research also aims at the other side to solving the issue of improving the removal efficiency of heavy metals from industrial wastewater by developing a low-cost treatment system for heavy metal removal from industrial wastewater. This system will be using a combination of an anaerobic sulfate reducing reactor with a macrophyte stabilization pond.

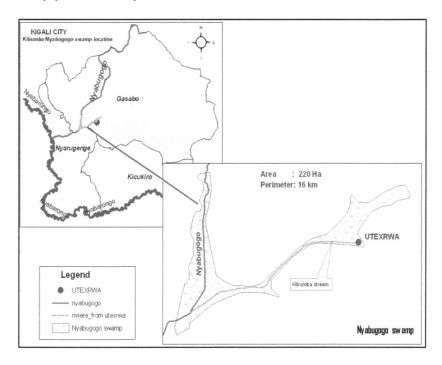

Figure1-1: Location of the textile industry (Utexrwa) and the Nyabugogo wetland in Kigali city.

1.4. Heavy metal removal from wastewater

Many processes exist for removing dissolved heavy metals, including ion exchange, precipitation, phytoextraction, ultrafiltration, reverse osmosis, and electrodialysis (Applegate, 1984; Sengupta and Clifford, 1986; Geselbarcht, 1996; Schnoor, 1997). Conventional biological and chemical treatment methods are not cost effective for heavy metal removal, because of the non degradability properties of heavy metals. These processes may be very expensive, especially when the metals in solution are in the concentration range of 1-100 mg/L (Nourbakhsh et al. 1994). Another disadvantage of conventional treatment technologies is the production of toxic chemical sludges of which the disposal / treatment is rather costly. Therefore, the wastewater contaminated by heavy metals need an effective and affordable technological solution.

Wetland plants are able to accumulate heavy metals whereas the content in the roots is considerably higher in comparison to the above-ground plant matter (Dushenko et al., 1995; Stoltz and Greger, 2002). However, their major shortcomings include:

> Saturation of metal bonding sites in the sediment (Machemer and Wildeman, 1992);
> Metal accumulation and uptake by plants play a minor role in wetlands and ponds systems (Mays and Edwards, 2001).

This study addresses the above-mentioned issues. We opted for affordable technologies for wastewater treatment. Waste stabilization ponds and wetlands systems will be applied because of their cost effectiveness, easy maintenance and no need of skilled personnel when compared to various advanced systems for treatment of wastewater that have been developed in EU and US and which requires high capital investments, maintenance costs and skilled personnel (Veenstra and Alaerts, 1996). It is also important to point out that in developing countries population and urbanization is on the rise (Gijzen and Khondker, 1997). Thus, important measures and actions need to be taken in order to manage efficiently the pollution problem in the near future.

Natural systems such as waste stabilization ponds, wetlands, land and grass treatment are the natural water purifier that also clean up the water in an eco-friendly way (Kootthatep and Polprasert, 2003). Wetland systems have been successfully used for water quality improvement. These act as reservoirs, part of wastewater treatment technology and also help to maintain the natural and biogeochemical links between land, water and biota. They also serve as available natural resource for wildlife and people (Connor and Luczak, 2002).

The bacterial processes in wetlands and especially in the rhizosphere can influence the fate of the metals. In the anaerobic zones in the presence of sulfate ions, the action of sulfate-reducing bacteria will mediate the fixation and precipitation of metals in the form of metal sulfides (Newman et al., 1997). Alternatively this provides possible mechanisms for improvement of heavy metals removal via precipitation. The problem of saturation of sediment bonding sites in anaerobic treatment has been dealt with introducing gravel in the system. The gravel is known to have adsorption capability and also to provide dissolved iron which plays an important role in metal co-precipitation in wetlands (Buddhawong et al., 2005). However, saturation of sediment bonding sites is a common problem of these systems.

1.5. Research Objectives

1.5.1. General objective

This research is aimed at the improvement of the mechanisms involved in the removal of heavy metals in the natural systems. The study addresses the issue of improving the removal efficiency of heavy metal from industrial wastewater in developing a low cost and alternative treatment system for metal removal from textile wastewater. That system consists of the combination of a bioreactor [Upflow Anaerobic Packed Bed Reactor (UAPBR)] with a macrophyte stabilization pond. The first step consists in the anaerobic treatment of wastewater. In this step, a new adsorbent tested for heavy metal removal will be used as filling material and surface attachment for bacterial growth. It is expected to remove a large quantity of heavy metals via adsorption combined to metal sulfide precipitation. The second step of the treatment will consist in phytoremediation where the metal will be removed via biosorption, uptake and translocation by macrophytes plants. The result of the above two systems will show how the two systems can react in a complementary way in removing heavy metals.

1.5.2. Specific objectives

The specific objectives of this research are:
1. To conduct a screening study of selected heavy metals (Cd, Cr, Cu, Pb and Zn) present in textile wastewater in an urban natural wetland case of the Nyabugogo natural wetland, receiving textile wastewater from UTEXRWA,
2. To determine the adsorption capacity, the effect of pH, the adsorption kinetic and the effect of the adsorbent dosage onto the removal of selected heavy metals on a volcanic rock,
3. To determine the heavy metal removal capacity of a pond system seeded with algae or duckweed, to show the effect of dissolved oxygen, pH and redox potential on the heavy metal removal efficiency in order to select a post treatment technology,
4. To develop an integrated cheap alternative for metal removal combining an Upflow Anaerobic Packed Bed Reactor (UAPBR) together with a macrophyte pond seeded with water hyacinth plants. The UAPBR contains the rock and it is using sulfate reduction.

1.6. Outline of the thesis

This thesis has been divided into seven chapters. The current chapter gives a background to the dissertation including the problems associated with the use of conventional and natural systems for wastewater treatment. Furthermore, it gives the objectives to be addressed; the relevance and the expected output of the study.

Chapter 2 presents a literature review on the use of natural systems (specifically wetlands) and bioreactor technologies for wastewater treatment.

Chapter 3 covers the screening of the water flowing in and out a Natural Wetland case of the Nyabugogo wetland located in Kigali city (Rwanda). This chapter focuses especially on heavy metal pollution coming from a textile factory and the industrial park entering the wetland and the fate of these metals within the system.

Chapter 4 discusses the removal of heavy metals looking at abiotic (dissolved oxygen, pH, redox potential and temperature) differences between two different types of pond treatments, i.e. algal versus duckweed plants.

Chapter 5 presents an investigation into the adsorption of heavy metals (Cd, Cu, Pb and Zn) onto volcanic rock from the Northern region of Rwanda as new adsorbent material for heavy metal removal from polluted water.

Chapter 6 develops an integrated system combining a sulfate reducing Upflow Anaerobic Packed Bed Reactor (UAPBR) with a water hyacinth pond as affordable and ecofriendly technique for the treatment of heavy metal polluted textile water. The UAPBR were filled with volcanic rocks tested in chapter 5 as adsorbent and attachement surface for bacterial growth.

Finally, Chapter 7 summarizes the results of this study and gives recommendations for practice and further research.

1.7. References

Al-Degs, Y., Khraisheh, M.A.M., Alen, S.J. & Ahmad, M.N. (2000). Effect of carbon surface chemistry on the removal of reactive dyes from textile effluent, Water Research 31(3), 927 – 935.

Applegate, L.E. (1984). Membrane separation processes, Chemical Engineering 91, 64 – 89.

Brik, M., Schoeberl, P., Chamam, B., Braun, R. & Fuchs, W. (2006). Advanced treatment of textile wastewater towards reuse using a membrane bioreactor. Process Biochemistry 41, 1751–1757.

Buddhawong, S., Kuschk, P., Mattusch, J., Wiessner, A. & Stottmeister, U. (2005). Removal of Arsenic and Zinc using different laboratory model wetland systems. Engineering Life Sciences. 5 (3), 247 – 252.

Chagas, E.P. & Durrant, L.R. (2001). Decolorization of azo dyes by *Phanerochaete chrysosporium* and *Pleurotus sajorcaju*. Enzyme and Microbial Technology 29, 473 – 477.

Cheng, S.P., Grosse, W., Karrenbrock, F. & Thoennessen, M. (2002). Efficiency of constructed wetlands in decontamination of water polluted by heavy metals. Ecological Engineering 18, 317 – 325.

Connor M.A. & Luczak A. (2002). Designing wetlands treatment systems that contribute to wildlife conservation. 8th international IWA Conference on wetland systems for water pollution control. Arusha, Tanzania, 16–19 Sept. 2002, 1024 – 1037.

Dushenko, W.T., Bright, D.A. & Reimer, K.J. (1995). Arsenic bio-accumulation and toxicity in aquatic macrophytes exposed to gold-mine effluent: Relationships with environmental partitioning, metal uptake and nutrients, Aquatic Botany 50, 141 – 158

Eckenfender, W.W. (1989). Industrial Water Pollution Control, McGraw-Hill Book Company, New York, NY.

Fenglian, F. & Qi, W. (2011). Removal of heavy metal ions from wastewaters: A review. Journal of Environmental Management 92, 407 – 418

Geselbarcht, J. (1996). Micro Filtration / Reverse Osmosis Pilot Trials for Livermore, California, Advanced Water Reclamation, in: Water Reuse Conference Proceedings, WWA 187.

Gijzen, H.J. & Khondker, M. (1997). An overview of ecology, physiology, cultivation and application of duckweed, Literature review. Report of Duckweed Research project. Dhaka, Bangladesh.

IFC, & world Bank group, (2007). Environmental, Health, and Safety Guidelines: Textile manufacturing. www.ifc.org/ifcext/enviro.nsf/Content/EnvironmentalGuidelines. (Accessed - June 2011)

Kadlec, R.H. & Wallace, S.D. (2008). Treatment Wetlands second edition, CRC Press.

Kootthatep T. & Polprasert C. (2003). Integrated pond/wetland and pond/aquaculture systems. In: Pond treatment technology (Ed. A. Shilton) Integrated Environmental Technology series, International water Association, London.

Machemer S.D. & Wildeman T.R. (1992). Adsorption compared to sulfide precipitation as metal removal processes from acid mine drainage in a constructed wetland, Journal of contaminant hydrology, 9, 115 – 131.

Mays P.A. & Edwards G.S. (2001). Comparison of heavy metal accumulation in a natural wetland and constructed wetlands receiving acid mine drainage, Ecological Engineering. 16, 487 – 500.

Merzouk, B., Gourich, B., Sekki, A., Madani, K. & Chibane, M. (2009). Removal turbidity and separation of heavy metals using electrocoagulation – electroflotation technique A case study. Journal of Hazardous Materials 164, 215 – 222.

MINAGRI, (2007). Ministry of Agriculture and Animal Resources. Good practice of agriculture in Rwanda, Kigali.

MINECOFIN, (2003). Ministry of Finance and Economic Planning. General Census of Population and Housing. Report on the Preliminary Results. National Census Service, Kigali.

MINIRENA, (2011). Ministry of Natural Resources. National policy for water resources management, Kigali.

MINITERE, (2004a). Ministry of Lands, Environment Forests. Water and Natural Resources, National Environmental Policy, Kigali.

MINITERE, (2004b). Ministry of Lands, Environment Forests. Water and Natural Resources, Sectorial policy on water and sanitation, Kigali.

Mohan, S.V., Prasad, K.K., Rao, N.C. & Sarma, P.N. (2005). Acid azo dye degradation by free and immobilized horseradish peroxidase (HRP) catalyzed process. Chemosphere 58, 1097 – 1105.

Moreira, M.T., Palma, C., Mielgo, I., Feijoo, G. & Lema, J.M. (2001). In vitro degradation of a polymeric dye (Poly R-478) by manganese peroxidase. Biotechnology and Bioengineering 75 (3), 362 – 368.

Newman D.K., Beveridge T.J. & Morel F.M.M. (1997), Precipitation of arsenic trisulfide by Desulfotomaculum auripigmentum, Applied and Environmental Microbiology 63, 2022 – 2028

Nourbakhsh, M., Sag, Y., Ozer, D., Aksu, Z., Katsal, T. & Calgar, A. (1994). Comparative study of various biosorbents for removal of chromium (VI) ions from industrial wastewater. Process Biochemistry 29, 1 – 5.

Oeko-Tex Association, International Association for Research and Testing in the Field of Textile Ecology. (2006a). Oeko-Tex Standard 100. Available at http://www.oeko-tex.com/en/main.html. (Accessed - June 2011)

Oeko-Tex Association, International Association for Research and Testing in the Field of Textile Ecology. (2006b). Oeko-Tex Standard 200. Available at http://www.oeko-tex.com/en/main.html. (Accessed - June 2011)

Oeko-Tex Association, International Association for Research and Testing in the Field of Textile Ecology. (2006c). Oeko-Tex Standard 1000. Available at http://www.oeko-tex.com/en/main.html. (Accessed - June 2011)

Official Gazette of the Republic of Rwanda, Organic Law N° 04 / 2005 of 08 / 04 / 2005 determining the modalities of protection, conservation and promotion of environment in Rwanda.

Schnoor, J.L. (1997). Phytoremediation, Technology Evaluation Report TE-97-01, Ground-Water Remediation Technologies Analysis Center, Pittsburgh, PA, USA.

Sengupta, A. K., & Clifford, D., (1986). Important process variables in chromate ion exchange, Environmental Science and Technology 20, 149 – 155.

Stoltz E. & Greger M. (2002), Accumulation properties of As, Cd, Cu, Pb, and Zn by four wetland plant species growing on submerged mine tailings. *Environ. Experim. Bot. 47*, 271 – 280.

Toh,Y.C., Yen, J.J.L., Obbard, J.P. & Ting,Y.P. (2003). Decolourisation of azo dyes by white-rot fungi (WRF) isolated in Singapore. Enzyme and Microbial Technology 33, 569 – 575.

Tsezos, M. (2001). Biosorption of metals. The experience accumulated and the 564 outlook for technology development. Hydrometallurgy 59, 241 – 243.

Veenstra, S. & Alaerts, G. (1996). Technology selection for pollution control. In: A.Balkema, H. Aalbers and E. Heijndermans (Eds.), Workshop on sustainable municipal waste water treatment systems, Leusdan, the Netherlands. 17 - 40.

Chapter 2: Treatment methods for textile wastewaters

Main part of this chapter has been submitted as:

Sekomo, C.B., Pakshirajan, K., Rousseau, D.P.L., & Lens, P.N.L., (2012). Perspectives on textile wastewater treatment using physico-chemical and biological methods including constructed wetlands. Submitted to the Journal of Chemical Technology and Biotechnology.

Abstract

Textile wastewater generally contains various pollutants, which can cause problems during the treatment process. These wastewaters have been for a long time treated using biological and physico-chemical methods. Due to the composition of textile wastewater, biological methods (activated sludge and anaerobic reactors) are suitable for their treatment. However, the recalcitrant and toxic nature of some dyes makes biological methods not efficient. Therefore, to increase the treatment efficiency recalcitrant dyes need to broken down by physico-chemical pre-treatment methods [advanced oxidation like Fenton's reagent (H_2O_2 + Fe^{2+}), hydrogen peroxide, ozonation and electro-coagulation] into easily biodegradable component. The use of natural and constructed wetlands is another method that can be used to treat textile wastewaters. This technique is cost effective and environmentally friendly. It uses phytoremediation techniques for cleaning up all kind of pollutants. It should be noted that there is no single technique which meets ideally all technical requirements. The final choice of a technology depends on many parameters like the efficiency of the technology, the pollutant to be treated, the investment and operating costs, the availability of local resources and the skilled personnel.

Keywords: Biological methods, dyes, heavy metals, natural system, physico-chemical methods.

2.1. Introduction

The textile industry is one of the oldest in the world. Its effluents are in general colored and represent severe environmental problems as they contain mixtures of chemicals and dyestuffs from different kinds. This makes the environmental challenge for textile industrial wastewater not only as a liquid waste, but also its varying chemical composition (Venceslau et al., 1994). The quality of textile wastewater depends very much on the employed coloring matters, dyestuffs and accompanying chemicals as well as the process itself. Depending on the season and the fashion, the composition of textile wastewater even of the same process changes often. About 8000 different coloring matters and 6900 additives are known and lead to an organic as well as an inorganic pollution of the wastewater (Brik et al., 2006). Due to the complexity of the textile processing, the wastewater generated contains usually elevated organic parameters such as chemical oxygen demand, total organic carbon, adsorbable organic halogens, inorganic compounds such as heavy metals, chloride, sulphate, sulphide and nitrogen ions. Unfixed dye releases large doses of color to the end of pipe effluents. Biological treatment processes are frequently used to treat textile effluents. These processes are generally efficient for biochemical oxygen demand (BOD_5) and suspended solids (SS) removal, but they are largely ineffective for removing color which is visible even at low concentrations (Slokar and Majcen, 1997; Banat et al., 1996).

Conventional treatment methods for textile wastewater are mainly physico-chemical (Kim et al., 2002; Kim et al., 2004; Golob et al., 2005) or biological (Kim et al., 2002; Chang et al., 2002; Libra et al., 2004). Conventionally, effluents containing organics are treated by adsorption, biological oxidation, coagulation, etc. Though the conventional methods have individual advantages, they are lacking of effectiveness if applied individually. For example, biological treatment is the most efficient and economic way of reducing the environmental impact of the industrial effluents containing organic pollutants, but this technique is time consuming and cannot be employed for textile effluent, as textile effluent contains many compounds that are recalcitrant to biodegradation. On the other hand, physical adsorption techniques are expensive and difficult for adsorbent regeneration. Further, biological and chemical methods generate considerable quantities of sludge, which itself requires treatment. Due to the large variability of the composition of textile wastewater, most of the traditional methods are becoming inadequate (Hao et al., 2000; Fernandes et al., 2004; Sakalis et al., 2005). Actually many technologies are available for treating wastewater from the textile industry, like: 1) adsorption, 2) biological treatment, 3) chemical precipitation, 4) electrochemical methods, 5) oxidation with ozone and 6) ultrafiltration (Sen and Demirer, 2003; Kurbus et al., 2003; Karcher et al., 2001; Ojstrsek et al., 2007; Golob and Ojstrsek, 2005). Compared to chemical treatment, biological treatment methods are cheaper than other methods, but dye toxicity usually inhibits bacterial growth and limits therefore the efficiency of the method (Greaves et al., 1999). Choosing the most appropriate treatment method or combination depends on the nature and amount of effluent from the textile processing plant. In this chapter, we review existing technologies for textile wastewater treatment currently available for this kind of wastewater generated.

2.2. Characteristics of textile wastewater

Textile wastewater effluents are related to wet operations, which are conducted during different parts of the textile manufacturing process. Wastewater from textile industry is typically alkaline and has high BOD and COD loads. Pollutants in textile wastewaters contain suspended solids, mineral oils (e.g. antifoaming agents, grease, spinning lubricant, non or low biodegradable surfactants [alkylphenol ethoxylates APEO, nonylphenol ethoxylates], and other organic compounds, including phenols from the wet finishing processes (e.g. dyeing), and halogenated organics from solvent used in bleaching (Brik et al., 2006; Kadlec and Wallace, 2008). The different operational steps in the textile industry are:

The scouring step involves the use of hot water and detergents to remove soil, vegetable impurities, grease (lanolin) and other contaminants from fibers. Wool scouring typically uses water and alkali, although scouring with an organic solvent is also possible. Scouring with alkali breaks down natural oils and surfactants and suspends impurities in the bath. The scouring effluent is strongly alkaline, and a significant portion of BOD_5 and COD loads from textile manufacturing arises from scouring processes (IFC and World Bank, 2007).

The desizing step generates effluents with significant concentrations of organic matter and solids. BOD_5 and COD loads from desizing may be significant (35 to 50 percent of the total load), and COD concentrations up to 20 g/L may be generated. Common bleaching reagents include hydrogen peroxide, sodium hypochlorite, sodium chlorite, and sulfur dioxide gas. Hydrogen peroxide is the most commonly used bleaching agent for cotton and is typically used with alkali solutions. The use of chlorine-based bleaches may produce organic halogens (due to secondary reactions) and cause significant concentrations of adsorbable organic halogens (AOX), particularly trichloromethane, in the wastewater. Sodium hypochlorite bleaching represents the most significant concern, and a lower AOX formation should result when using sodium chlorite bleaching. The wastewater is alkaline (IFC & world Bank group, 2007).

A mercerizing step where the cotton fiber reacts with a solution of caustic soda, and a hot-water wash treatment that removes the caustic solution from the fiber. The caustic solution remaining on the fiber is neutralized with acid, followed by a number of rinses to remove the acid. Wastewater from mercerizing is highly alkaline, since it contains caustic soda. The recommended pollution prevention and control technique involves the recovery and reuse of alkali from mercerizing effluent, particularly rinsing water, subject to color limitations that may apply to mercerized cloth woven from dyed yarn (IFC & world Bank group, 2007).

The dyeing step containing color pigments, halogens (especially in vat, disperse, and reactive dyes), metals (e.g. copper, chromium, zinc, cobalt, and nickel), amines (produced by azo dyes under reducing conditions) in spent dyes, and other chemicals used as auxiliaries in dye formulation (e.g. dispersing and antifoaming agents) and in the dyeing process (e.g. alkalis, salts, and reducing / oxidizing agents). Dyeing process effluents are characterized by relatively high BOD and COD values, the latter commonly above 5 g/l. Salt concentration (e.g. from reactive dye use) may range between 2 - 3 g/l (IFC & world Bank group, 2007),

The printing step where components consist of color concentrates, solvents, and binder resins. Color concentrates contain pigments (insoluble particles) or dyes. Organic solvents are used exclusively with pigments. Defoamers and resins are aimed at increasing color fastness. Printing blankets or back grays (fabric backing material that absorbs excess print paste), which are washed with water before drying, may generate wastewater with an oily appearance and significant volatile organic compound (VOC) levels from the solvents (mineral spirits) used in print paste (IFC & world Bank group, 2007).

The mothproofing step is based on the use of biocides like permethrin, cyfluthrin and other, which are potentially highly toxic compounds to aquatic life.

Since textile manufacturing operations use a myriad of raw materials, chemicals and processes, wastewater treatment may require the use of unit operations specific to the manufacturing process in use. Techniques for treating industrial process wastewater in this sector include source segregation and pretreatment of wastewater streams as follows: (i) high load (COD) streams containing non-biodegradable compounds using chemical oxidation, (ii) reduction in heavy metals using chemical precipitation, coagulation and flocculation, etc. and (iii) treatment of highly colored or high TDS streams using reverse osmosis (IFC & world Bank group, 2007). Additional engineering controls may be required for (i) advanced metals removal using membrane filtration or other physical/chemical treatment technologies (Sen and Demirer, 2003), (ii) removal of recalcitrant organics, residual pesticides and halogenated organics using activated carbon or advanced chemical oxidation (Ayse, 1997; Morias and Zamora, 2005; Kobya et al., 2006; Mohan et al., 2007; Merzouk et al., 2009; Belkacem et al., 2008), (iii) residual color removal using adsorption or chemical oxidation, (iv) reduction in effluent toxicity using appropriate technology (such as reverse osmosis, ion exchange, activated carbon, etc.), (v) reduction in TDS in the effluent using reverse osmosis or evaporation, and (vi) containment and neutralization of nuisance odors (Karcher et al., 2001; Golob and Ojstrsek, 2005; Ojstrsek et al., 2007).

2.3. Biological methods for textile wastewater

Biological treatment of wastewater is often the most cost-effective method when compared to other treatment options (Morias and Zamora, 2005). Biological methods are generally cheap, simple to apply and they have been applied to remove organics and color compounds of textile wastewater (Kim et al. 2007). However, the conventional aerobic biological process, e.g., activated sludge process, cannot readily treat textile wastewater, because most commercial dyes are toxic to the organisms used in the process (Kim et al., 2002; Koch et al., 2002).

2.3.1. Aerobic and anaerobic bioreactors

In order to avoid toxicity problems and to improve the treatment efficiencies, biological methods are used in combination with physico-chemical methods. The combination of activated sludge units and membrane filtration for biomass retention generally results in high effluent qualities and compact plant configurations (Brik et al., 2006). Complete solids removal, a significant disinfection capability, high rate and high efficiency organic removal and small footprint are common requirements for every wastewater treatment system. An important feature of membrane bioreactors is the possibility to employ high sludge ages facilitating the growth of specialized microorganisms and in such a way promoting improved degradation of refractory organics (Stephenson et al., 2000). This makes the membrane bioreactor technology a highly promising technique for industrial wastewater purification. Therefore, membrane bioreactor effluents can be of a quality suitable for direct recycling or after further purification by additional post-treatment steps. It was demonstrated that the system is largely resistant to changing loading rates and that even at high loading rates efficient COD removal occurs. The apparent sludge yield was very low underlining an additional advantage of the membrane bioreactor systems. With regard to maximum color removal, sludge growth was found to be of critical importance (Malpei et al., 2003; Brik et al., 2004). Due to the persistent nature of textile dyes, low sludge ages should be employed, to generate sufficient new biomass allowing to adsorb incoming color loads. Despite the superior performance compared to other biological treatment systems, recalcitrant COD and color components are still present in the effluent to an extent unacceptable for direct reuse purposes. Nevertheless, even if wastewater reclamation is intended the membrane bioreactor technology is the method of choice when it is combined with other advanced treatment technologies.

The recalcitrant nature of azo dyes, together with their toxicity to microorganisms, makes aerobic treatment difficult. A wide range of azo dyes can be decolorized anaerobically (Sen and Demirer, 2003; Razo-Flores et al. 1997, Banat et al. 1996, Brown and Laboureur, 1983, Carliell et al. 1995, Chinwekitvanich et al. 2000). Under anaerobic conditions, azo dyes are readily cleaved via a four-electron reduction at the azo linkage generating aromatic amines. The required electrons are provided by electron donating carbon source such as starch, volatile fatty acids or glucose. In addition, it is known that methanogenic and acetogenic bacteria in anaerobic microbial consortia contain unique reduced enzyme cofactors, such as F 430 and vitamin B_{12} that could also potentially reduce azo bonds (Razo-Flores et al. 1996, Zaoyan et al. 1992). These steps remove the color of the dye; however, they do not completely mineralize the aromatic amines generated in the anaerobic environment (Brown and Laboureur, 1983; Kool, 1984; Zeyer et al. 1985) with a few exceptions (Razo-Flores et al. 1996, O'Connor and Young, 1993). As suspect mutagens and carcinogens, the aromatic amines cannot be regarded as environmentally safe end products. It is known that most of the aromatic amines can be biodegraded under aerobic conditions (Carliell et al., 1995; Brown and Hamburger, 1987; Seshadri et al., 1994). Several high rate anaerobic reactors have been developed for treating wastewaters at relatively short hydraulic retention times (24 hours). Of these the anaerobic fluidized bed reactor (FBR) has been one of the technological advances (Sen and Demirer, 2003). It has been successfully employed in a broad spectrum of wastewaters including both readily and hardly biodegradable wastes (Denac and Dunn, 1988; Henze and Harremoes, 1983; Hickey and Owens, 1981). Besides, several high rate anaerobic reactors such as anaerobic baffled reactors (ABR) (Bell et al., 2000), upflow anaerobic sludge blanket reactors (UASB) (Razo-Flores et al., 1997; Chinwekityanich et al., 2000; Huren et al., 1994; Donlon et al., 1997; O'Neill et al., 2000) and upflow anaerobic packed bed reactors (UAPR) Sekomo et al., (2011b); were used in the anaerobic treatment of synthetic or real textile wastewater.

Heavy metals are found among the pollutants present in textile wastewaters. One of the best available technologies for the removal of metals from wastewater is the application of sulphate-reducing bacteria (SRB). This process is based on the production of biogenic hydrogen sulphide by SRB, which consequently reacts with metal ions in the water forming sparingly soluble metal sulfides. Retention of heavy metals has been achieved efficiently, with > 90 % metal removal efficiencies (La et al., 2003) achieved using Upflow Anaerobic Packed Bed reactors (UAPB) and Upflow Anaerobic Sludge Blanket reactors (UASB) as reported by Jong and Parry (2003, 2006), Lens et al. (2002), Van Hullebusch et al. (2006, 2007), Sekomo et al. (2011b). Precipitation of metals, e.g. as sulfides or carbonates, is usually reported to be the main removal mechanism. The reduction of sulphate to sulfides by bacteria enhances dramatically the heavy metal removal efficiency compared to biosorption mechanisms due to the very low solubility of metal sulfides (Sekomo et al., 2011b, Jong and Parry, 2003; La et al., 2003; Quan et al., 2003). However, it is important to stress that due to the complexity of the wastewater composition, i.e. the presence of refractory organics, heavy metals, and reducing agents, inhibition of sensitive microbial groups and insufficient removal in bioreactors may occur in some cases. As reported by Correira et al. (1994) textile wastewater is composed by a diverse kind of pollutant depending on the step in the manufacturing process and this has a particular impact on the biological treatment (Table 2-1).

Table 2-1: Pollutants types in textile wastewaters, their origin and impact in biological treatment (modified after Correira et al., 1994).

Pollutants	Chemical types	Processes of origin	Impact on biological treatment
Color	Dyes, scoured wool impurities	Dyeing Scouring	Insufficient removal in bioreactor
Nutrients	Ammonium salts, urea, phosphate based buffers and sequestrants	Dyeing	No removal in anaerobic processes, biological nutrient removal required
Organic load	Enzymes, fats, greases, starches, surfactants, waxes	Desizing Dyeing Scouring Washing	High demand on aeration systems activated sludge bulking problems
pH	Carbonates, mineral and organic acids, silicate, sodium hydroxide, sodium chloride	Bleaching Desizing Dyeing Mercerizing Neutralization Scouring	Inhibition and collapse of bioreactors
Refractory organics	Carrier organic solvents, chlorinated organic compounds, dyes, resins, surfactants	Bleaching Desizing Dyeing Finishing Scouring Washing	Insufficient removal in bioreactors, possible accumulation in biomass aggregates / films leading to inhibition
Sulphur	Sulphuric acid, sulphate, sulphide, hydrosulphite salts	Dyeing	Sulphate reduction in anaerobic reactors
Toxicants	Heavy metals, oxidizing and reducing agents, peroxide, dichromate biocides	Bleaching Desizing Dyeing Finishing	Inhibition of sensitive microbial groups (methanogens, nitrifiers) in bioreactors

2.3.2. Natural and constructed wetlands

For more than four decades, it has been proved experimentally that aquatic plants have potentiality in removing contaminants from aquatic environments (Wolverton and Mckown, 1976, Brix, 1987; Kadlec and Knight, 1996; Rousseau et al., 2004, Sangeeta and Savita, 2009, Sekomo et al., 2011a). The main objective in the wastewater treatment process is to eliminate or reduce contaminants to the level that causes no adverse effect on humans or the receiving environment. Natural and constructed wetlands are appreciated over conventional methods as they are ecological friendly, require less energy inputs and hence are economical.

Macrophytes plants play a prominent role in nutrient and heavy metal recycling of many aquatic eco-systems (Pip and Stepaniuk, 1992). Heavy metals and other contaminants can be removed by microorganisms or by aquatic plants. Aquatic plants are commonly observed in water bodies throughout the world (Reddy, 1984). A water body enriched with nutrients either by natural processes or by urban and agricultural runoff supports the luxuriant growth of aquatic plants, algae etc. Aquatic plants are suitable for wastewater treatment because they have tremendous capacity of absorbing nutrients and other substances from the water (Boyd, 1970b) and hence bring the pollution load down. Denny (1980, 1987) further noted that the main route of heavy metal uptake in aquatic plants was through the roots in the case of emergent and surface floating plants, whereas in submerged plants both roots and leaves take part in removing heavy metals and nutrients. Cowgill (1974) suggested that submerged rooted plants have potential from water as well as sediments, whereas rootless plants extract metals rapidly only from the water phase.

The macrophytes commonly found in eutrophic water bodies include:
- Free floating macrophytes (entire plant body except the roots is above water), e.g. *Eichhornia crassipes*, *Ludwigia sp.*, *Salvinia,* etc.
- Submerged macrophytes (whole plant body remaining submerged in water), e.g. *Hydrilla sp.*, *Egeria,* etc.
- Emergent macrophytes (plants rooted in soil but emerging to significant heights above the water), e.g. *Typha sp.*, *Phragmites sp.*

2.3.2.1. Potential of different macrophytes plants in improving water quality

A) Free floating macrophytes (Duckweed and Water Hyacinth)

Duckweed is a small green plant that belongs to the *Lemnacae* family. They are the smallest flowering plants, which grow floating on still or slow-moving fresh or brackish water. The duckweed family has four genera: *Lemna, Spirodela, Wolffia* and *Wolfiella* with 37 species (Gijzen and Khondker, 1997). Their growth is usually associated with nutrient-rich water. This is the reason why they have been employed in wastewater treatment. Various studies have shown that this plant has the potential of taking up nutrients (Gijzen and Khondker, 1997; Al-Nozaily et al., 2000a; Al-Nozaily et al., 2000b; Zimmo et al., 2000; Smith and Moelyowati, 2001; Cheng et al., 2002; El-Shafai, 2004; Nhapi I., 2004; Caicedo, 2005) and heavy metals (Duong, 2001; Meggo, 2001, Rousseau et al., 2011; Sekomo et al., 2012b) from wastewater.

The duckweed plant has been used in stabilization pond systems for wastewater treatment, which represent an important class of wastewater treatment systems in developing countries because of their cost-effectiveness (Gijzen, 2001b). Waste stabilization ponds are low cost wastewater treatment systems producing high-quality effluents that allow water reuse in irrigation (Zimmo et al. 2004). Different authors have proposed the use of duckweed ponds for the efficient and low-cost treatment of domestic and industrial wastewater at urban or rural levels (Diederik et al., 2011; Shi et al., 2010; Dalu and Ndamba 2003). As reported by Diederik et al. (2011), duckweed has a high metal uptake capacity specifically for Cr, Zn and Pb. As they did not investigate the metal partitioning in the plant, their removal is assumed as biosorption combined with uptake.

Water hyacinth or *Eichhornia crassipes* is one of the most troublesome aquatic weeds in the world (Holm et al., 1977). Water hyacinth has been studied for its tendency to accumulate the heavy metal contaminants present in water bodies. Mitchell (1976) reported that its population can double in less than a week. This plant has been used in removing heavy metals and nutrients. Lot of work had been compiled on the efficiency and mechanism of removing contaminants through this macrophyte, due to its tendency to bio-accumulate and biomagnify the heavy metal contaminants present in water bodies (Mishra and tripati, 2009; Hassan et al., 2007; Soltan and Rashed, 2003). Salati (1987) reported a study on heavy metal uptake by water hyacinth in Brazil and observed that it is a plant with good tolerance and high uptake of nutrients and heavy metals. Muramoto and Oki (1983) found water hyacinth capable of accumulating pollutants in their study on bioaccumulation of heavy (Pb, Cr, Cu, Cd, and Zn) from contaminated water. Tiwari et al. (2007) explained that heavy metals Pb, Zn, Mn show greater affinity towards bioaccumulation in their study the presence of higher concentrations of heavy metals in plants signifies the biomagnification. Water hyacinth has the unique property to accumulate heavy metals Cd, Cu, Pb and Zn from the root tissues of the plant (Soltan and Rashed, 2003; Nor, 1990).

The rapid growth of water hyacinth on polluted water was reported by Wolverton and McDonald (1976), this is an advantage for its capacity for purification and adsorption of heavy metals. Furthermore, they also reported a decrease in Biological Oxygen Demand (BOD) of polluted waters by the same plant. The efficiency is due to the absorption of the organic matter by the root of the water hyacinth. These roots act as filters through mechanical and biological activity, removing suspended particles from the water and decreasing turbidity. Wolverton (1989), Brix (1993) and Johnston (1993) explained that reason for turbidity reduction, i.e. the root hairs have electrical charges that attract opposite charges of colloidal particles such as suspended solids and cause them to adhere on the roots where they are slowly digested and assimilated by the plant and microorganisms. Brix (1993) reported that water hyacinth was successfully used in wastewater treatment systems, therefore improving the water quality by reducing the levels of inorganic and organic pollutants. Stowell et al. (1981) reported water hyacinth lagoons functioning as horizontal trickling filters in which submerged plant roots provide physical support for a thick bacterial biofilm. This actually degrades organic matter. Moorhead and Reddy (1988) studied that release from plant root appears to be a major source of oxygen in the lagoon.

B) Submerged Macrophytes (*Hydrilla verticillata, Salvinia*)

The whole plant is submerged in water and thus the whole plant plays on important role in the removal of contaminants. Denny (1980) observed that the reliance upon roots for heavy metal uptake was in rooted floating-leaved taxa with lesser reliance in submerged taxa. He also observed that tendency to use shoots as sites of heavy metal uptake instead of roots increases with progression towards submergence and simplicity of shoot structure. Scientists are also carrying out studies for removal of contaminants through *Hydrilla verticillata* in the last few years. Elankumaran et al., (2003) established a comparative study between *Hydrilla verticillata* and *Salvinia sp.* and concluded that the removal efficiency of *Hydrilla verticillata* is higher at lower (at 5 ppm) concentration of copper compared to *Salvinia*, but in higher concentrations the removal efficiency of *Salvinia* is higher. They also observed that morphology was dependent upon the duration of exposure and the initial dose, i.e. morphological changes were observed at longer exposure and higher dose. The removal of heavy metal ions by the nonliving biomass of aquatic macrophytes was studied by Bunluesin et al. (2007). They found that *Hydrilla verticillata* is an excellent biosorbent for treating wastewater with a low concentration of Cd contaminants.

Algae are microscopic organisms that are cosmopolitan in distribution. They exist in fresh water, oceans, ice, hot springs etc. They are rich in diversity and have many functions ranging from medicinal and food to wastewater treatment. In wastewater treatment, they are mostly employed in oxidation ponds. In the presence of light, algae produce oxygen via a process called photosynthesis. This is fundamental in promoting bacterial conversion of organic matter and nitrogenous substances (Fruend et al., 1993; Mara et al., 1992) as well as photo-oxidation of pathogens (Curtis et al., 1992a). Algae are known to be natural systems for wastewater treatment (Fruend et al., 1993). The major limitation they have is essentially due to the lack of an elaborate root system, which presents a specific area for bacterial attachment and growth. The attachment of algae to biofilms increases the pH values due to the effect of oxygen production or carbon dioxide consumption in the system.

C) Emergent Macrophytes (*Phragmities australis, Phragmities carka, Typha latifolia*)

Emergent macrophytes are generally found at the bank of rivers and lakes. Roots are attached in the soil and play an important role in taking up heavy metals and nutrients from the soil sediment. Emergent macrophytes have more supportive tissues than floating macrophytes. They might have a greater potential for storing the nutrients over a longer period. Juwarker et al. (1995) have reported 78% to 91% removal of BOD, nitrogen reduction from 30.8 to 9.8 mg/l and phosphate reduction from 14.9 to 9.6 mg/l using the emergent macrophytes *Typha latiofolia* and *Phragmites carka*. Emergent plants influence metal storage indirectly by modifying the substratum through oxygenation, buffering, and pH and adding organic matter (Dunabin and Bowmer, 1992). *Phragmities australis* have been observed in removing heavy metal from wastewater (Du Laing et al., 2003). They have reported on different analytical digestion method applicable for heavy metal (Cd, Cu Mn, Ni Pb and Zn) in leaves and stem of the plant. Their analysis results showed the potential of this plant in metal removal.

Constructed wetlands are systems simple to use, environmentally friendly, with low construction and operational costs, and efficient enough to treat diverse wastewaters, although the experience in treating textile wastewaters is limited (Figure 2-1). Constructed wetland's designs differ regarding to the type of flow and the filter material used (Reed, 1993; Kadlec and Knight, 1996, Naz et al., 2009). For many decades the technology of wastewater treatment by means of constructed wetlands or engineered systems with horizontal sub-surface flow was started in Germany based on research by Käthe Seidel in the 1960s (Seidel, 1961, 1964, 1966) and by Reinhold Kickuth in the 1970s (Kickuth, 1977, 1978). In these systems, the wastewater is fed via the inlet and flows slowly through the porous medium under the surface of the bed in a more or less horizontal path until it reaches the outlet zone where it is collected before leaving via a level control arrangement at the outlet. During this passage, the wastewater will come into contact with a network of aerobic, anoxic and anaerobic zones. The aerobic zones occur around roots and rhizomes that leak oxygen into the substrate (Brix, 1987; Cooper et al., 1996).

For engineered systems, major design parameters, removal mechanisms and treatment performance have been reviewed by many researchers Kadlec et al. (2000), Vymazal (2005), Vymazal and Kröpfelová (2008), Kadlec and Wallace (2008) and Vymazal

(2009). Treatment behaviour in CWs is often considered to be a figurative black-box (Rousseau et al., 2004). Detailed understanding of constructed wetland functioning is still desirable because a large number of physical, chemical and biological processes occur in parallel and influence each other. Until now, constructed wetland design has been mainly based on rules of thumb approaches using specific surface area of requirements (Brix and Johansen, 2004) or simple first-order decay models (Kadlec and Knight, 1996; Rousseau et al., 2004).

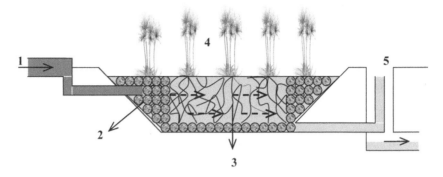

Figure 2-1: Representation of an engineered wetland with horizontal sub-surface flow. Where 1 represent the inlet pipe, 2 the distribution zone filled with large stones; 3 medium (filled with gravel and sand); 4, vegetation and 5 the outlet pipe.

The use of engineered systems imitating the self-cleaning ability of natural wetland ecosystems by establishing optimal physical, chemical and biological conditions for in situ wastewater treatment should be considered with greater importance (Tjasa and Alenka, 2008; Scholz and Xu, 2002). Mbuligwe (2005) reported that different types of wetland plants may have relative treatment efficiency advantages. Color removal from dye-rich wastewater was 72 – 77 % in the vegetated constructed wetland units and only 14 % in the unplanted units. COD removal efficiencies were 68 – 73 % in the vegetated units and 51 % in the control unit. Sulphate removal was 53 % in the vegetated units while in the unplanted unit it was only 15 %.

Furthermore, as reported by Vymazal (2009), the use of horizontal flow constructed wetlands for treatment of textile wastewaters were carried out as early as during late 1980s and early 1990s in Germany and Australia. Colored wastewater, with high COD (up to 35 g/L), TSS (up to 25 g/L) and pH values and low BOD_5 values (BOD_5/COD ratio usually between 0.1 and 0.4), is treated satisfactorily in horizontal flow constructed wetlands with very good removal of COD, TSS, ammonia, sulfate and anionic sulfate. Also, the removal of visible colorization is high. On the other hand, low BOD/COD ratios indicate the hardly-degradable nature of textile wastewater and therefore, high BOD_5 removal efficiencies cannot be expected. Even if in some case no high BOD removal efficiency could be reached, constructed wetland as such is a very promising method and cost-effective option for colored textile wastewater.

2.4. Physico-chemical methods

As environmental regulations become stringent, new and novel processes for efficient treatment of various kinds of wastewater at relatively low operating cost are needed. In this context, researchers are trying various alternative processes, such as electrochemical techniques, wet oxidation, ozonization, photocatalytic methods for the degradation of organic compounds. Among these advanced oxidation processes, the electrochemical treatment has been receiving greater attention in recent years due to its unique features, such as versatility, energy efficiency, automation and cost effectiveness (Gutierrez and Crespi, 1999; Lorimer et al., 2001; Mohan et al., 2007; Merzouk et al., 2010; Belkacem et al., 2008).

A) Electrochemical techniques

In electrochemical techniques, the main reagent is the electron, called "Clean Reagent", which degrades all the organics present in the effluent without generating any secondary pollutant or by-product / sludge. Electrochemical techniques offer high removal efficiencies and have lower temperature requirements compared to non-electrochemical treatment. In addition to the operating parameters, the rate of pollutant degradation depends of the anode material. When electrochemical reactors operate at high cell potential, the anodic process occurs in the potential region of water discharge, hydroxyl radicals are generated (Simond et al., 1997). On the other hand, if chloride is present in the electrolyte, an indirect oxidation via active chlorine can be operative (Kotz et al., 1991; Szpyrkowicz et al., 2005). Naumczyk et al. (1996) have demonstrated several anode materials, such as graphite and noble metal anodes can be successfully used for the oxidation of organic pollutants. Comninellis (1994) experimented anodic oxidation of phenol in the presence of NaCl using a tin oxide coated titanium anode and reported second order kinetics for the degradation of phenol at the electrode surface. Fernandes et al. (2004) studied the degradation of C.I. Acid Orange 7 using boron-doped diamond electrode and reported that more than 90 % COD removal. Anastasios et al. (2005) demonstrated 94 % dye removal using a pilot plant electrochemical reactor for textile wastewater treatment.

Recently, investigations have focused on the treatment of wastewaters using electrocoagulation (EC). The electrocoagulation has been successfully used for the treatment of wastewaters such as electroplating wastewater (Adhoum et al., 2004), laundry wastewater (Ge et al., 2004), latex particles (Vorobiev et al., 2003), restaurant wastewater (Chen et al., 2000) and slaughterhouse wastewater (Kobya et al., 2006). Meanwhile, the EC process has been widely used in the removal of arsenic (Kumar et al., 2004), phosphate (Bektas et al., 2004), sulfide, sulfate and sulfite (Murugananthan et al., 2004), boron (Yilmaz et al., 2005), fluoride (Koparal and Ogutveren, 2002), nitrate (Gao et al., 2004) and chromate (Pons et al., 2005). Treatment of textile wastewaters by electrocoagulation showed that COD, color, turbidity and dissolved solids at varying operating conditions are considerably removed (Kobya et al., 2004;). In addition, it is clear that a technically efficient process must also be economically feasible with regard to its initial capital and operating costs, and practical to be applied to environmental problems. The economic aspect of the electrocoagulation process is not well investigated (Kobya et al., 2004). Electrical energy consumption is a very important economical parameter in the electrocoagulation process, like in all other electrolytic processes (Merzouk et al., 2009).

B) Oxidative techniques

The oxidative process is a convenient technique currently being used in the industry for decolorization treatments of wastewater, for sterilization of water and for the biological treatment of water. These oxidative processes for waste water treatment comprise: biological oxidation; oxidation with UV irradiation; oxidation with sodium hypochlorite, hydrogen peroxide, ozone and nitric acid; incineration and wet oxidation (low pressure / high pressure). In the oxidative process, the dyes in the water do not have to be fully decomposed for decolorization to occur. As the dyes are oxidized, they are broken down into small, colorless molecules. It also destroys dyes which contain metals. Although this technique decolorizes the water, the metals in phthalocyanine, metal complexes and formazan complex dyes are released in the form of free cobalt, nickel, copper and chromium ions and this technique is unable to remove the metal ions (Ayse, 1997). Dyes also contain some chemicals such as nitrogen, chlorine or sulphur. The products of these molecules following the oxidation process may make them more toxic than the parent molecule. There is insufficient knowledge about the relative pollution load and the toxicity of the treated wastewater which contains these chemicals (Ayse, 1997).

Generally, organic compounds can be degraded by advanced oxidation processes. These have been used to enhance the biodegradability of wastewaters containing refractory and / or non-biodegradable organic substances which can be toxic to microorganisms (Lin and Peng, 1994; Vlyssides and Papaioannou, 2000; Kim et al., 2004; Kim et al., 2007). Morias and Zamora (2005) studied two advanced oxidation processes (Fe^{2+} / H_2O_2 / UV and H_2O_2 / UV systems) to enhance the biodegradability of landfill leachates. At optimized experimental conditions (2000 mg/L of H_2O_2 and 10 mg/L of Fe^{2+} for the photo-Fenton system, and 3000 mg/L of H_2O_2 for the H_2O_2 / UV system), both methods showed suitability for partial removal of chemical oxygen demand (COD), total organic carbon (TOC) and color. The biodegradability was significantly improved (BOD_5 / COD from 0.13 to 0.37 or 0.42) which allowed an almost total removal of COD and color by a sequential activated sludge process. The application of an integrated system of electron beam radiation and biological treatment to textile wastewater can be a powerful process. The electron beam radiation can transform refractory organic compounds into easily biodegradable products, thus improving the efficiency and reducing the cost of a further biological treatment step.

C) Evaluation

Physico-chemical techniques are promising but they need further improvement. These techniques should be applied prior to using biological treatment because their reaction products are biodegradable. An ideal technique is the one which is easily applicable to all kind of wastewaters, has low economical running and capital costs, is applicable to a large volume of wastewater, does not require a large area, would leave no sludge, disinfects the water to a degree that it is reusable and finally not harmful to the environment. However, it should be noted that there is no single technique which meets these ideal requirements. As reported by Ayse (1997), all these techniques have both advantages and disadvantages as overviewed in Table 2-2.

Table 2-2: Summary of physico-chemical decolorization method (Ayse, 1997)

Method	Comments
Sodium hypochlorite	Effective on decolorisation, cheaper than other oxidants, and easily applicable (20 – 40°C, 5 – 30 min). Risk of halogenated hydrocarbon (AOX) increase and bacterial toxicity. Can only be used with small amounts of wastewater.
Hydrogen peroxide	Environmentally friendly application. Not effective on all dyes as oxidation potential is not very high.
Fenton's reagent	More effective than hydrogen peroxide on different classes of dyes. Waste water may be reused following this treatment and removes heavy metals. Causes severe sludge problems.
Ozone	Specially useful in decolorisation of water-soluble dyes. Does not sufficiently decrease COD and turbidity. Acids, aldehydes and ketones are reaction products. Recommended that coagulation and ozone can be used prior to biological treatment.
UV irradiation	Photocatalytic reactions of some organic species in aqueous solutions are feasible. Removes heavy metals. Sludge and harmful UV scattering problems.
Gamma irradiation	New technique.
UV irradiation hydrogen peroxide	Increased rate and strength of oxidation, but the cost of producing UV irradiation does not compensate for this increase. Environmentally friendly application, apart from some UV scattering.
UV irradiation and ozone	Increased rate and strength of oxidation, but the cost incurred by the UV irradiation does not compensate for this increase. Environmentally friendly, apart from UV and ozone scattering. Waste water may be reused since reaction products could be carbon dioxide, water, nitrogen, etc.

2.5. Conclusions

Aside from the aesthetic deterioration and obstruction for penetration of dissolved oxygen and light into the natural water bodies caused by the presence of color and suspended particles from dyes, dye degradation products are carcinogenic and mutagenic in nature. Biological treatment methods provide efficient and low cost means of textile wastewater treatment. Furthermore, bioreactor technologies have been used to treat textile wastewater under aerobic and anaerobic conditions. Treatment of textile wastewater under aerobic condition was not efficient in pollutant removal due to dyes toxicity. Treatment of textile wastewater under anaerobic conditions was found to be optimal. Existing physico-chemical methods for decolorization such as advanced oxidation processes like the use of Fenton's reagent ($H_2O_2 + Fe^{2+}$), hydrogen peroxide and ozonation are expensive and generate sludge problems. The coagulation–flocculation technique produces also high amounts of sludge, causing handling and disposal problems. Activated carbon adsorption, membrane filtration, irradiation, and electro kinetic coagulation are also expensive and not very established. It should be kept in mind that anaerobic treatment compared to physico-chemical treatment methods still offers a viable option in terms of cost. In contrast, the use of natural and constructed wetlands as a low cost and ecofriendly wastewater treatment option for developing countries has demonstrated well its capability to treat dye-rich textile wastewater. The results coming from this natural system indicate that reduction of pollution parameters depends mostly on the chemical composition of textile wastewater and the design of the system.

2.6. References

Adhoum, N., Monser, L., Bellakhal, N. & Belgaied, J.   (2004). Treatment of electroplating wastewater containing Cu^{2+}, Zn^{2+} and Cr(VI) by electrocoagulation, Journal of Hazardous Materials B 112, 207 – 213.

Al-Nozaily, F., Alaerts, G. & Veenstra, S. (2000a). Performance of duckweed–covered sewage lagoon. 1. Oxygen balance and COD removal. Water Research 34 (10), 2727 – 2733.

Al-Nozaily, F., Alaerts, G. & Veenstra, S. (2000b). Performance of duckweed–covered sewage lagoon. 2. Nitrogen and phosphorus balance and plant productivity. Water Research 34 (10), 2734 – 2741.

An, H., Qian, Y., Gu, X.S. & Tang, W.Z. (1996). Biological treatment of dye wastewaters using an anaerobic-oxic system. Chemosphere 33, 2533 – 2542.

Anastasios, S., Konstantinos, M., Ulrich, N., Konstantinos, F. & Anastasios, V. (2005). Evaluation of a novel electrochemical pilot plant process for azo dyes removal from textile wastewater, Chemical Engineering Journal 111, 63 – 70.

Ayse, U. (1997). An overview of oxidative and photooxidative decolorization treatments of textile wastewaters, Journal of the Society of Dyers and Colourists 113, 211 – 217.

Banat, I.M., Nigam, P., Singh, D. & Marchant, R. (1996). Microbial decolorization of textile-dye-containing effluents: a review. Bioresource Technology 58, 217 – 227.

Bektas, N., Akbulut, H., Inan, H. & Dimoglu, A. (2004). Removal of phosphate from aqueous solutions by electrocoagulation, Journal of Hazardous Materials B 106, 101 – 105.

Belkacem, M., Khodir, M. & Abdelkrim, S. (2008). Treatment characteristics of textile wastewater and removal of heavy metals using the electroflotation technique. Desalination 228, 245 – 254.

Bell, J., Plumb, J.J., Buckley, C.A. & Stuckey, D.C. (2000). Treatment and decolorization of dyes in an anaerobic baffled reactor. Journal of Environmetal Engineering Div ASCE 126, 1026 – 1032.

Boyd, C.E. (1970b). Production, mineral nutrient accumulation and pigment concentration in Typha latifolia and Scripus americaus. Ecology, 51, 285–290.

Brik, M., Chamam, B., Schoberl, P., Braun, R. & Fuchs, W. (2004). Effect of ozone, chlorine and hydrogen peroxide on the elimination of colour in treated textile wastewater by MBR. Water Science and Technology 49 (4), 299 – 303.

Brik, M., Schoeberl, P., Chamam, B., Braun, R. & Fuchs, W. (2006). Advanced treatment of textile wastewater towards reuse using a membrane bioreactor. Process Biochemistry 41, 1751–1757.

Brix, H. (1993). Macrophytes -mediated oxygen transfer in wetlands: Transport mechanism and rate. In G. A. Moshiri (Ed.), Constructed wetlands for water quality improvement. Ann Arbor, London: Lewis.

Brix, H. (1987). Treatment of wastewater in the rhizosphere of wetland plants – the root zone method. Water Science and Technology 19: 107 – 118.

Brix, H. & Johansen, N.H. (2004). Guidelines for Vertical Flow Constructed Wetland System up to 30 PE. No. 52. Denmark, Copenhagen.

Brown, D. & Hamburger, B. (1987). The degradation of dye stuffs: 3 Investigations of their ultimate degradability. Chemosphere 16, 1539 – 1553.

Brown, D. & Laboureur, P. (1983). The degradation of dyestuffs: 1. Primary biodegradation under anaerobic conditions. Chemosphere 12, 397 – 404.

Bunluesin, S., Krutrachue, M., Pokethitiyook, P., Upatham, S. & Lanza, G.R. (2007). Batch and continuous packed column studies of cadmium biosorption by Hydrilla verticillata biomass. Journal of Bioscience and Bioengineering, 103(6), 509 – 513.

Caicedo B.J.R. (2005) Effect of operational variables on nitrogen transformations in duckweed stabilization ponds, PhD thesis, IHE Delft The Netherlands

Carliell, C.M., Barclay, S.J., Naidoo, N., Buckley, C.A., Mulholland, D.A. & Senior, E. (1995). Microbial decolorization of a reactive azo-dye under anaerobic conditions. Water SA 21, 61 – 69.

Chang, W.S., Hong, S.W. & Park, J. (2002). Effect of zeolite media for the treatment of textile wastewater in a biological aerated filter. Process Biochemistry 37, 693 – 698.

Chen, X., Chen, G. & Yue, P.L. (2000).Separation of pollutants from restaurant wastewater by electrocoagulation, Separation and Purification Technology 19, 65 – 76.

Cheng, J., Bergamann, B.A., Classen, J.J., Stomp, A.M. & Howard, J.W. (2002). Nutrient recovery from swine lagoon water by spirodela punctata. Bioresource Technology 81, 81–85

Chinwekitvanich, S., Tuntoolvest, M. & Panswad, T. (2000). Anaerobic decolorization of reactive dye bath effluents by a two stage UASB system with tapioca as a co-substrate. Water Research 34, 2223 – 2232.

Comninellis, C. (1994). Electrocatalysis in the electrochemical conversion/ combustion of organic pollutants for wastewater treatment, Electrochimica Acta 39 (11/12), 1857–1863.

Cooper, P.F., Job, G.D., Green, M.B. & Shutes, R.B.E. (1996). Reed Beds and Constructed Wetlands for Wastewater Treatment. WRc Publications, Medmenham, Marlow, UK.

Correia, V.M., Stephenson, T. & Judd, S.J. (1994). Characterization of textile wastewaters. A review. Environmental Technology, 15, 917 – 929.

Cowgill, V.M. (1974). The hydro geochemical of Linsley Pond, North Braford. Part 2. The chemical composition of the aquatic macrophytes. Archiv fur Hydrobiologie, 45(1), 1–119.

Curtis, T.P., Mara, D.D. & Silva, A.S. (1992a). Influence of pH, Oxygen, and Humic substances on the ability of sunlight to damage fecal coliforms in waste stabilization pond water. Applied Environmental Microbiology 58(4), 1335 – 1343.

Denac, M. & Dunn, I.J. (1988). Packed-bed and fluidized bed biofilm reactor performance for anaerobic wastewater treatment. Biotechnology and Bioengineering 32, 159 – 173.

Denny, P. (1980). Solute movement in submerged angiosperms. Biological Review, 55, 65–92.

Denny, P. (1987). Mineral cycling by wetland plants a review. Archiv fur Hydrobiologie Beith, 27, 1–25.

Donlon, B., Razo-Flores, E., Luijten, M., Swarts, H., Lettinga, G. & Field, J. (1997). Detoxification and partial mineralization of the azo dye mordant orange 1 in a continuous upflow anaerobic sludge-blanket reactor. Applied Microbiology and Biotechnology 47, 83 – 90.

Du Laing, G.., Tack, F.M.G. & Verloo, M.G. (2003). Performance of selected destruction methods for the determination of heavy metals in reed plants (Phragmities australis). Analytica Chimica Acta, 497(1–2), 191 – 198.

Dunabin, J.S. & Bowmer, K.H. (1992). Potential use of constructed wetlands for treatment of industrial wastewater containing metals. Science of the Total Environment, 3, 151–168.

Duong, H.D. (2001) The use of aquatic plants for the removal of heavy metals from water, MSc. Thesis, Gent University Belgium

Elankumaran, R., Raj, M.B. & Madhyastha, M.N. (2003). Biosorption of copper from contaminated water by Hydrilla verticillata Casp. and Salvinia sp. Green Pages, Environmental News Sources.

El–Shafai M. (2004) Nutrients valorization via duckweed wastewater treatment and aquaculture, PhD thesis, IHE Delft The Netherlands

Fernandes, A., Morao, M., Magrinho, A., Lopes, I. & Goncalves, (2004). Electrochemical degradation of C.I. Acid Orange 7. Dyes Pigments 61, 287–297.

Fruend, C., Romem, E. & Post, A.F. (1993). Ecological physiology of an assembly of photosynthetic micro algae in wastewater oxidation ponds. Water Science and Technology 27(7-8), 143-149.

Gao, P., Chen, X., Shen, F. & Chen, G. (2004). Removal of chromium(VI) from wastewater by combined electrocoagulation–electroflotation without a filter, Separation and Purification Technology 43, 117 – 123.

Ge, J., Qu, J., Lei, P. & Liu, H. (2004). New bipolar electrocoagulation–electroflotation process for the treatment of laundry wastewater, Separation and Purification Technology 36, 33 – 39.

Gijzen H.J. (2001b). Low cost wastewater treatment and potentials for re-use: a cleaner production approach to wastewater management. In: proceedings Intl. Symposium on low cost wastewater treatment and re-use. Cairo, Egypt, February 3-4.

Gijzen, H.J. & Khondker, M. (1997). An overview of ecology, physiology, cultivation and application of duckweed, Literature review. Report of Duckweed Research project. Dhaka, Bangladesh.

Golob, V. & Ojstrsek, A. (2005). Removal of vat and disperse dyes from residual pad liquors, Dyes pigments 64, 57 – 61.

Golob, V., Vinder, A. & Simonic, M. (2005). Efficiency of the coagulation/flocculation method for the treatment of dyebath effluents. Dyes Pigments 67, 93 – 97.

Greaves, A.J., Phillips, D.A.S. & Taylor, J.A. (1999). Literature Review, Journal of the Society of Dyers and Colourists 115, 363 – 365.

Gutierrez, M.C. & Crespi, M. (1999). A review of electrochemical treatments for colour elimination, Journal of the Society of Dyers and Colourists 115 342 – 345.

Hao, O.J., Kim, H. & Chiang, P.C. (2000). Decolorization of wastewater. Critical review. Environmental Science & Technology 30 (4), 449 – 505.

Hassan, S.H., Talat, M. & Rai, S. (2007). Sorption of zinc and cadmium from aqueous solutions by water hyacinth (Eichhornia crassipes), Bioresource Technology 98, 918–928.

Haug, W., Schmidt, A., Nortemann, B., Hempel, D.C., Stolz, A. & Knackmuss, H.J. (1991). Mineralization of the sulfonatedazo dye mordant yellow-3 by a 6-aminonaphthalene-2-sulfonatedegrading bacterial consortium. Applied Environmental Microbiology 57, 3144 – 3149.

Henze, M. & Harremoes, P. (1983). Anaerobic treatment of wastewater in fixed film reactors - a literature review. Water Science and Technology 15, 1–101.

Hickey, R.F. & Owens, R.W. (1981). Methane generation from high strength industrial wastes with the anaerobic biological fluidized bed, Biotechnology and Bioengineering Symposium 11, 399 – 413.

Holm, L.G., Plucknett, D.L., Pancho, V. & Herberger, J.P. (1977). The world's worst weeds: Distribution and biology p. 609. Honolulu: University Press of Hawaii.

Huren, A., Yi, Q. & Xiasheng, G. (1994). A way for water pollution control in dye manufacturing industry. 49th Purdue Industrial Waste Conference Proceedings, Chelsea, MI, USA: Lewis Publishers.

IFC, & world Bank group, (2007). Environmental, Health, and Safety Guidelines: Textile manufacturing. www.ifc.org/ifcext/enviro.nsf/Content/EnvironmentalGuidelines. (Accessed - June 2011)

Johnston, C.A. (1993). Mechanism of water wetland water quality interaction. In G. A. Moshiri (Ed.), Constructed wetland for water quality improvement (pp. 293–299). Ann Arbor: Lewis.

Jong, T. & Parry, D.L. (2003). Removal of sulphate and heavy metals by sulphate reducing bacteria in short-term bench scale upflow anaerobic packed bed reactor runs. Water Research 27, 3379 – 3389.

Jong, T. & Parry, D.L. (2006). Microbial sulphate reduction under sequentially acidic conditions in an upflow anaerobic packed bed bioreactor, Water Research 40(13), 2561 – 2571.

Juwarker, A.S., Oke, B., Juwarkar, A. & Patnaik, S.M. (1995). Domestic wastewater treatment through constructed wetland in India. Water Science Technology, London, 32(3), 291–294.

Kadlec, R.H. & Knight, R.L. (1996). Treatment Wetlands, Lewis Publishers, CRC Press, Inc., Boca Raton, FL, USA.

Kadlec, R.H., Knight, R.L., Vymazal, J., Brix, H., Cooper, P. & Haberl, R. (2000). Constructed wetlands for pollution control. In: Processes, Performance, Design and Operation. IWA Specialist Group on the Use of Macrophytes in Water Pollution Control, IWA Scientific and Technical Report N°. 8, IWA Publishing, London.

Kadlec, R.H. & Wallace, S.D. (2008). Treatment Wetlands, 2nd ed. CRC Press, Boca Raton, FL.

Karcher, S., Kornmuller, A. & Jekel, M. (2001). Screening of commercial sorbents for the removal of reactive dyes, Dyes Pigments 5, 111 – 125.

Kickuth, R. (1977). Degradation and incorporation of nutrients from rural wastewaters by plant hydrosphere under limnic conditions. In: Utilization of Manure by Land Spreading, Comm. Europ. Commun., EUR 5672e, London, 335 – 343.

Kickuth, R. (1978). Elimination gelöster Laststoffe durch Röhrichtbestände. Arbeiten des Deutschen Fischereiverbandes 25, 57 – 70.

Kim, T., Park, C., Lee, J., Shin, E. & Kim, S. (2002). Pilot scale treatment of textile wastewater by combined process (fluidized biofilm process – chemical coagulation – electrochemical oxidation). Water Research 36, 3979 – 3988.

Kim, T., Park, C., Yang, J. & Kim, S. (2004). Comparison of disperse and reactive dye removals by chemical coagulation and Fenton oxidation. Journal of Hazardous Materials 112, 95 – 103.

Kobya, M., Can, O.T., Bayramoglu, M. & Sozbir, M. (2004). Operating cost analysis of electrocoagulation of textile dye wastewater, Separation and Purification Technology 37, 117 – 125.

Kobya, M., Senturk, E. & Bayramoglu, M. (2006). Treatment of poultry slaughterhouse wastewaters by electrocoagulation, Journal of Hazardous Materials B 133, 172 – 176.

Koch, M., Yediler, A., Lienert, D., Insel, G. & Kettrup, A. (2002). Ozonation of hydrolyzed azo dye reactive yellow 84 (CI). Chemosphere 46, 109–113.

Kool, H.J. (1984). Influence of microbial biomass on the biodegradability of organic compounds. Chemosphere 13, 751 – 761.

Koparal, A.S. & Ogutveren, U.B. (2002). Removal of nitrate fromwater by electroreduction and electrocoagulation, Journal of Hazardous Materials B 89, 83 – 94.

Kotz, R., Stucki, S. & Carcer, B. (1991). Electrochemical wastewater treatment using high over voltage anodes. Part I. Physical and electrochemical properties of SnO_2 anodes, Journal of Applied Electrochemistry 21, 14 – 20.

Kuai, L., De Vreese, I., Vandevivere, P. & Verstraete, W. (1998). GA Camended UASB reactor for the stable treatment of toxic textile wastewater. Environmental Technology 19, 1111 – 1117.

Kumar, P.R., Chaudhari, S., Khilar, K.C. & Mahajan, S.P. (2004). Removal of arsenic from water by electrocoagulation, Chemosphere 55, 1245 – 1252.

Kurbus, T., Majcen Le Marechal, A. & Brodnjak Voncina, D. (2003). Comparison of H_2O_2/UV, H_2O_2/O_3 and H_2O_2/Fe^{2+} processes for the decolorization of vinylsulphone reactive dyes, Dyes Pigments 58, 245 – 252.

La, H.J., Kim, K.H., Quan, Z.X., Cho, Y.G. & Lee, S.T. (2003). Enhancement of sulphate reduction activity using granular sludge in anaerobic treatment of acid mine drainage. Biotechnology Letters 25, 503 – 508.

Lens, P.N.L., Vallero, M., Esposito, G. & Zandvoort, M. (2002). Perspectives of sulphate reducing bioreactors in environmental biotechnology. Reviews in Environmental Science and Biotechnology. 1 (4), 311 – 325.

Libra, J.A., Borchert, M., Vigelahn, L. & Storm, T. (2004). Two stage biological treatment of a diazo reactive textile dye and the fate of the dye metabolites. Chemosphere 56, 167 – 180.

Lin, S.H. & Peng, C.F. (1994). Treatment of textile wastewater by electrochemical method. Water Research. 28, 277 – 283.

Lorimer, J.P., Mason, T.J., Plattes, M., Phull, S.S. & Walton, D.J. (2001). Degradation of dye effluent, Pure and Applied Chemistry 73 (12) 1957 – 1968.

Lourenco, N.D., Novais, J.M. & Pinheiro, H.M. (2000). Reactive textile dye color removal in a sequencing batch reactor. Water Science and Technology 42, 321 – 328.

Malpei, F., Bonomo, L. & Rozzi, A. (2003). Feasibility study to upgrade a textile wastewater treatment plant by a hollow fibre membrane reactor for effluent reuse. Water Science and Technology 47(10 – 11), 33 – 39.

Manu, B. & Chaudhari, S. (2002). Anaerobic decolorisation of simulatedtextile wastewater containing azo dyes. Bioresource Technology. 82, 225 – 231.

Mara, D.D., Alabster, G.P., Pearson, H.W & Mills, S.W. (1992). Waste stabilization ponds, a design manual for Eastern Africa, Lagoon Technology International Leeds, England

Mbuligwe, S.E. (2005). Comparative treatment of dye-rich wastewater in engineered wetland systems (EWSs) vegetated with different plants, Water Research 39: 271 – 280.

Meggo R. (2001) Allocation pattern of Lead and Zinc in Cyperus papyrus and Lemna gibba, MSc. Thesis, IHE Delft The Netherlands

Merzouk, B., Gourich, B., Sekki, A., Madani, K. & Chibane, M. (2009). Removal turbidity and separation of heavy metals using electrocoagulation – ectroflotation technique A case study. Journal of Hazardous Materials 164, 215 – 222.

Merzouk, B., Madani, K. & Sekki, A. (2010). Using electrocoagulation–electroflotation technology to treat synthetic solution and textile wastewater, two case studies. Desalination 250, 573 – 577.

Mishra, V.K. & Tripathi, B.D. (2009). Accumulation of chromium and zinc from aqueous solutions using water hyacinth (Eichhornia crassipes). Journal of Hazardous Materials 164, 1059 – 1063.

Mitchell, D. S. (1976). The growth and management of Eichhornia crassipes and Salvinia spp. in their native environment and in alien situations. In C. K. Varshney, & J. Rzoska (Eds.), Aquatic weeds in Southeast Asia (p. 396). The Hague: Dr. W. Junk.

Mohan, N., Balasubramanian, N. & Ahmed Basha, C. (2007). Electrochemical oxidation of textile wastewater and its reuse. Journal of Hazardous Materials 147, 644 – 651.

Moorhead, K.K. & Reddy, K.R. (1988). Oxygen transport through selected aquatic macrophytes. Journal of Environmental Quality 17(1), 138–142.

Morias, J.L. & Zamora, P.P. (2005). Use of advanced oxidation process to improve the biodegradability of mature landfill leachate. Journal of Hazardous Materials B123, 181–186.

Muramoto, S. & Oki, Y. (1983). Removal of some heavy metals from polluted water by water hyacinth (Eichhornia crassipes). Bulletin of Environmental Contamination and Toxicology, 30, 170–177.

Murugananthan, M., Raju, G.B. & Prabhakar, S. (2004). Removal of sulfide, sulphate and sulfite ions by electrocoagulation, Journal of Hazardous Materials B 109, 37 – 44.

Naumczyk, J., Szpyrkowicz, L., De Faveri, M. & Zilio Grandi, F. (1996). Electrochemical treatment of tannery wastewater containing high strength pollutants, Trans. I Chem. E 74 B, 58 – 59.

Naz, M., Uyanik, S., Yesilnacar, M.I. & Sahinkaya, E. (2009). Side-by-side comparison of horizontal subsurface flow and free water surface flow constructed wetlands and artificial neural network (ANN) modeling approach. Ecological Engineering 35, 1255 – 1263.

Nhapi I. (2004) Options for wastewater management in Harare, Zimbabwe, PhD thesis, IHE Delft The Netherlands

Nigam, P., Banat, I.M., Singh, D. & Marchant, R. (1996). Microbial process for the decolorization of textile effluent containing azo, diazo and reactive dyes. Process Biochemistry 31, 435 – 442.

Nor, Y.M. (1990). The absorption of metal ions by Eichhornia crassipes. Chemical Speciation and Bioavailability 2, 85 – 91.

O'Connor, O.A. & Young, L.Y. (1993). Effect of nitrogen limitation on the biodegradability and toxicity of nitrophenol and aminophenol isomers to methanogenesis. Archives of Environmental Contamination and Toxicology 25, 285 – 291.

O'Neill, C., Hawkes, F.R., Hawkes, D.L., Esteves, S. & Wilcox, S.J. (2000). Anaerobic-aerobic biotreatment of simulated textile effluent containing varied ratios of starch and azo dye. Water Research 34, 2355 – 2361.

Ojstrsek, A., Fakin, D. & Vrhovsek, D. (2007). Residual dyebath purification using a system of constructed wetland, Dyes Pigments 74, 503 – 507.

Oxspring, D.A., McMullan, G., Smyth, W.F. & Marchant, R. (1996). Decolourisation andmetabolism of the reactive textile dye, remazol black B, by an immobilized microbial consortium. Biotechnology Letters 18, 527 – 530.

Panswad, T. & Luangdilok, W. (2000). Decolorization of reactive dyes with different molecular structures under different environmental conditions. Water Research 34, 4177 – 4184.

Pip, E. & Stepaniuk, J. (1992). Cadmium, copper and lead in sediments. Archiv fur Hydrobilogie, 124, 337 – 355.

Pons, M.N., Alinsafi, A., Khemis, M., Leclerc, J.P., Yaacoubi, A., Benhammou, A. & Nejmeddine, A. (2005). Electrocoagulation of reactive textile dyes and textilewastewater, Chemical Engineering Processing. 44, 461 – 470.

Quan, Z.X., La, H.J., Cho, Y.G., Hwang, M.H., Kim, I.S. & Lee, (2003). Treatment of metal contaminated water and vertical distribution of metal precipitates in an upflow anaerobic bioreactor. Environmental Technology. 24, 369 – 376.

Razo-Flores, E., Donlon, B., Field, J. & Lettinga, G. (1996). Biodegradability of N-substituted aromatics and alkylphenols under methanogenic conditions using granular sludge. Water Science and Technology 33, 47 – 57.

Razo-Flores, E., Luijten, M., Donlon, B., Lettinga, G. & Field, J. (1997) Biodegradation of selected azo dyes under methanogenic conditions. Water Science and Technology 36, 65 – 72.

Reddy, (1984). Water hyacinth for water quality improvement and biomass production. Journal of Environmental Quality, 13, 1– 8.

Reed, S.C. (1993). Subsurface Flow Constructed Wetlands for Wastewater Treatment—A Technology Assessment, EPA.

Rousseau, D.P.L., Vanrollegham, P.A. & Pauw, N.D. (2004). Model based design of horizontal subsurface flow constructed treatment wetlands: a review. Water Research 38, 1484 – 1493.

Rousseau. D.P.L., Sekomo, C.B., Saleh, S.A.A.E. & Lens, P.N.L. (2011), Duckweed and Algae Ponds as a Post-Treatment for Metal Removal from Textile Wastewater. Water and nutrient management in natural and constructed wetlands. 63 – 75, DOI: 10.1007/978-90-481-9585-5_6.

Sakalis, A., Mpoulmpasakos, K., Nickel, U., Fytianos, K. & Voulgaropoulos, A. (2005). Evaluation of a novel electrochemical pilot plant process for azodyes removal of textile wastewater. Chemical Engineering Journal, 111 (1): 63 – 70.

Salati, E. (1987). Edaphic–phytodepuration: A new approach to waste water treatment. In K. R. Reddy, & W. H. Smith (Eds.), Aquatic plants for water treatment and resource recovery (pp. 199–208). Orlando Fl: Magnolia.

Sangeeta, D. & Savita, D. (2009). Water quality improvement through macrophytes : a review. Environmental Monitoring and Assessment 152, 149 – 153.

Scholz, M. & Xu, J. (2002). Performance comparison of experimental constructed wetlands with different filter media and macrophytes treating industrial wastewater contaminated with lead and copper, Bioresource Technology 83, 71 – 79.

Seidel, K. (1961). Zur Problematik der Keim- und Pflanzengewasser. Verh. Internat. Verein. Limnol. 14, 1035 – 1039.

Seidel, K. (1964). Abau von Bacterium coli durch höhere Pflanzen. Naturwissenschaften 51, 395.

Seidel, K. (1966). Reinigung von Gewässern durch höhere Pflanzen. Naturwissenschaften 53, 289 – 297.

Sekomo, C.B., Kagisha, V., Rousseau, D.P.L. & Lens, P.N.L. (2011b). Heavy metal removal by a combined system of anaerobic upflow packed bed reactor And water hyacinth pond. Environmental Technology. DOI:10.1080/09593330.2011.633564

Sekomo, C.B., Nkuranga, E., Rousseau, D.P.L. & Lens, P.N.L. (2011a). Fate of Heavy Metals in an Urban Natural Wetland: The Nyabugogo Swamp (Rwanda). Water Air and Soil Pollution 214, 321 – 333.

Sekomo, C.B., Rousseau, D.P.L., Saleh, S.A. & Lens, P.N.L. (2012b). Heavy metal removal in duckweed and algae ponds as a polishing step for textile wastewater treatment. Ecological Engineering 44, 102 - 110.

Sen, S. & Demirer, G.N. (2003). Anaerobic treatment of real textile wastewater with a fluidized bed reactor. Water Research 37, 1868 – 1878.

Seshadri, S., Bishop, P.L. & Agha, A.M. (1994). Anaerobic/aerobic treatment of selected azo dyes in wastewater. Waste Management 14, 127 – 137.

Shi, W., Wang, L., Rousseau, D.P.L. & Lens, P.N.L. (2010). Removal of 589 estrone, 17α-ethinylestradiol, and 17β-estradiol in algae and duckweed-based 590 wastewater treatment systems. Environmental Science and Pollution Research 17(4), 824-833.

Simond, O., Schaller, V. & Comninellis, C.H. (1997). Theoretical model for the anodic oxidation of organics on metal electrodes, Electrochim. Acta 42 (34) 2009 – 2012.

Slokar, Y.M. & Majcen, M.A. (1997). Methods of decolorization of textile wastewater. Dyes and Pigments 37 (4), 335 – 356.

Smith, M.D. & Moelyowati, I. (2001). Duckweed based wastewater treatment (DWWT): design guidelines for hot climates. Water Science and Technology. 43(11), 291–299

Soltan, M.E. & Rashed, M.N. (2003). Laboratory study on the survival of water hyacinth under several conditions of heavy metal concentrations. Advances in Environmental Research 7, 321 – 334.

Stephenson, T., Judd, S., Jefferson, B. & Brindle, K. (2000). Aerobic membrane bioreactors treating industrial wastewaters. In: Membrane bioreactors for wastewater treatment. London: IWA Publishing.

Stowell, R., Ludwig, R., Colt, J. & Tchobanoglous, T. (1981). Concepts in aquatic treatment design. Journal of Environmental Engineering, ASCE, 112, 885 – 894.

Szpyrkowicz, L., Santhosh, N.K., Neti, R.N. & Satyanarayan, S. (2005). Influence of anode material on electrochemical oxidation for the treatment of tannery wastewater, Water Research 39, 1601–1613.

Tiwari, S., Dixit, S. & Verma, N. (2007). An effective means of bio-filtration of heavy metal contaminated water bodies using aquatic weed Eichhornia crassipes. Environmental Monitoring and Assessment, 129, 253–256.

Tjasa, G. & Bulc, Alenka Ojstrsek, (2008). The use of constructed wetland for dye-rich textile wastewater treatment. Journal of Hazardous Materials 155, 76 – 82.

Van Hullebusch, E.D., Farges, F., Lenz, M., Lens, P.N.L. & Brown, G.E.J. (2007). Selenium speciation in biofilms from granular sludge bed reactors used for wastewater treatment. AIP. Conference Proceedings. 882, 229–231.

Van Hullebusch, E.D., Gieteling, J., Zhang, M., Zandvoort, M.H., Daele, W.V., Defrancq, J. & Lens, P.N.L. (2006). Cobalt sorption onto anaerobic granular sludge: isotherm and spatial localization analysis. Journal of Biotechnology 121 (2), 227–240.

Venceslau, M.C., Tom, S. & Simon, J.J. (1994). Characterization of textile wastewaters- a review. Environmental Technology, 15, 917 – 929.

Virendra Kumar Mishra, B.D. Tripathi (2009) Accumulation of chromium and zinc from aqueous solutions using water hyacinth (*Eichhornia crassipes*). Journal of Hazardous Materials 164, 1059 – 1063.

Vlyssides, A.G. & Papaioannou, D. (2000). Testing an electrochemical method for treatment of textile dye wastewater. Waste Management 20, 569 – 574.

Vorobiev, E., Larue, O., Vu, C. & Durand, B. (2003). Electrocoagulation and coagulation by iron of latex particles in aqueous suspensions, Separation and Purification Technology 31: 177 – 192.

Vymazal, J. (2005). Horizontal sub-surface flow and hybrid constructed wetland systems for wastewater treatment. Ecological Engineering 25, 478 – 490.

Vymazal, J. (2009), The use constructed wetlands with horizontal sub-surface flow for various types of wastewater. Ecological Engineering 35, 1 – 17.

Vymazal, J. & Kröpfelová, L. (2008). Wastewater Treatment in Constructed Wetlands with Horizontal Sub-Surface Flow. Springer, Dordrecht.

Wolverton, & McDonald, (1976). Don't waste waterweeds. New Scientist, 71, 318–320.

Wolverton, B.C. (1989). Aquatic plant/microbial filters for treating septic tank effluent in wastewater treatment. In D. A. Hammer (Ed.), Municipal industrial and agricultural waste. Chelsea MI: Lewis.

Wolverton, B.C. & Mckown, M. M. (1976). Water hyacinth for removal of phenols from polluted waters. Aquatic Botany, 30, 29 – 37.

Yilmaz, A.E., Boncukcuoglu, R., Kocakerim, M.M. & Keskinler, B. (2005). The investigation of parameters affecting boron removal by electrocoagulationmethod, Journal of Hazardous Material. B 125: 160 – 165.

Zaoyan, Y., Ke, S., Guangliang, S., Fan, Y., Jinshan, D. & Huanian, M. (1992). Anaerobic-aerobic treatment of a dye wastewater by combination of RBC with activated sludge. Water Science and Technology 26, 2093 – 2096.

Zeyer, J., Wasserfallen, A. & Timmis, K.N. (1985). Microbial mineralization of ring substituted anilines through an orto cleavage pathway. Applied and Environmental Microbiology 50 (2), 447 – 453.

Zimmo O.R., van der Steen N.P. & Gijzen, H.J. (2004) Nitrogen mass balance across pilot-scale algae and duckweed-based wastewater stabilisation ponds. Water Research 38, 913 – 920.

Chapter 3: Fate of Heavy Metals in an Urban Natural Wetland: The Nyabugogo Swamp (Rwanda)

This chapter was presented and published as:

Sekomo, C.B., Rousseau, D.P.L., & Lens, P.N.L., (2009). Fate of Heavy Metals in an Urban Natural Wetland: The Nyabugogo Swamp (Rwanda). In: *Proceedings of the WETPOL 2009: 3rd Wetland Pollutant Dynamics and Control.* Caixa Forum and Hotel Barcelona Plaza, Barcelona, Spain (20 – 24 September 2009).

Sekomo, C.B., Nkuranga, E., Rousseau, D.P.L., & Lens, P.N.L., (2011a). Fate of Heavy Metals in an Urban Natural Wetland: The Nyabugogo Swamp (Rwanda). Water Air and Soil Pollution 214, 321 – 333.

Abstract

The Nyabugogo natural wetland (Kigali City, Rwanda) receives all kinds of untreated wastewaters, including those from industrial areas. This study monitored heavy metal concentrations (Cd, Cr, Cu, Pb and Zn) in all environmental compartments of the swamp: water and sediment, the dominant plant species *Cyperus papyrus*, and fish (*Clarias* sp. and *Oreochromis* sp.) and Oligochaetes. Cr, Cu and Zn concentrations in the water were generally below the WHO (2008) drinking water standards whereas Cd and Pb were consistently above these limits. Except Cd, all metal concentrations were below the threshold levels for irrigation. The highest metal accumulation occurred in the sediment with up to 4.2 mg/kg for Cd, 68 mg/kg for Cu, 58.3 mg/kg for Pb and 188.0 mg/kg for Zn, followed by accumulation in the roots of *Cyperus papyrus* with up to 4.2 mg/kg for Cd, 45.8 mg/kg for Cr, 29.7 mg/kg for Cu and 56.1 mg/kg for Pb. Except Cu and Zn, other heavy metal (Cd, Cr and Pb) concentrations were high in *Clarias* sp., *Oreochromis* sp. and Oligochaetes. Therefore, there is a human health concern for people using water and products from the swamp.

Key words: *Cyperus papyrus*, industrial pollution, bioaccumulation, Nyabugogo wetland, heavy metal.

3.1. Introduction

People in developing countries have been using wetlands for water purification for quite a long time (Denny, 1987, Mitsch and Jørgensen, 1989, 2004; Blackwell et al., 2002; Zedler and Kercher, 2005; Verhoeven et al., 2006). Pollution occurring in wetlands depends on many factors, the most important ones being the increase of the population and the associated economical and industrial development. In Kigali city (Rwanda), the pollution of the Nyabugogo wetland has been identified as a potential threat. This is mainly caused by the fact that many industries do not have on site treatment plants. Therefore, they are discharging their wastewater in two small rivers, called Kibumba and Ruganwa, which discharge in the Nyabugogo swamp. This study is focused on heavy metals from industries because if the concentrations of these pollutants are not controlled, they will have an adverse impact on people using this water. Even if the appearance of water (color and taste) has helped in preventing people from using it for cooking or drinking, they are still using it for washing clothes, irrigating their crops and consume fish captured in the swamp.

Many researchers have published on the dynamics of heavy metals in wetlands with respect to vegetation and animals living there. Judith and Peddrick (2004) stressed the importance of knowing the processes of metal removal, uptake and distribution in the wetland. The extent of uptake and how metals are distributed within plants can have important effects on the residence time of metals in plants, in wetlands and the potential release of metals when conditions change. Knowing this, one could understand these systems and ensure that wetlands do not themselves become sources of metal contamination. When using wetlands for wastewater treatment, intentionally or not, it is necessary to be aware of the amounts of pollutant that can be sequestrated by the wetland. Ewers and Schlipkoter (1991), Denny et al. (1995) and Monday et al. (2003) reported on the negative impact of heavy metals where they enter the food chain and accumulate in fish tissues, particularly into the liver; and Fleming and Trevors (1989) reported that copper in its ionized form could be lethal to fish.

This chapter presents an investigation on the fate of heavy metals conducted in the Nyabugogo Swamp. This swamp receives wastewater from different sources, the most important ones being the textile industry (Utexrwa) and the Gikondo industrial park. Wastewaters from these different sources are dumped directly into the Nyabugogo swamp without any treatment. In order to assess the potential pollution, the present research has carried out a monitoring of heavy metal concentrations inside the Nyabugogo swamp and in the rivers discharging to and from the swamp. Also the fate of metals entering the swamp has been monitored through analyses of plants (*Cyperus papyrus*), sediment, fish (*Clarias* sp. and *Oreochromis* sp.) and invertebrates (Oligochaetes) living in the swamp.

3.2. Materials and methods

3.2.1. Study area

The Nyabugogo wetland is located in Kigali, the capital of Rwanda. This natural wetland receives wastewater from houses and industries (Figure 3-1). The Nyabugogo swamp is the part which is not exploited by the population for agriculture. Its surface area, according to CGIS (Centre of Geographic Information Systems, Butare, Rwanda), is 60.09 ha.

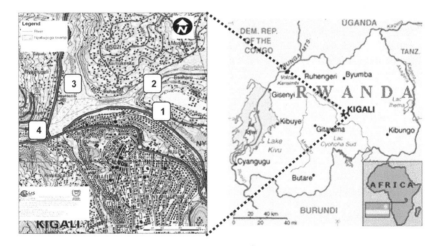

Figure 3-1: Location of the Nyabugogo swamp and its inlet rivers (1, 2 and 3) and outlet river (4). Source: Centre of Geographic Information Systems (CGIS, NUR / Butare, 2006)

3.2.2. Sampling sites

Monitoring of heavy metals in the Nyabugogo swamp was conducted in three different campaigns. Samples have been taken both in the rainy (April 07 and December 08), and dry (October 07) season. The procedure used consisted in partitioning the swamp in three different transects (1, 2 and 3) along which water samples were taken at three different locations (A, B and C), as shown in Figure 2. Plant and sediment samples were collected at only one point along each transect, i.e. 1A, 2A and 3B. Water samples were also collected in the rivers entering and exiting the swamp. In addition, two fish species (*Clarias* sp. and *Oreochromis* sp.) and macroinvertebrates (Oligochaetes) were caught to assess bioaccumulation of heavy metals. A total of thirteen sampling sites were thus established, two sites in the Nyabugogo River (numbers 3 & 4 in Figure 3-1), one site in the Kibumba River (number 2 in Figure 3-1), one site in the Ruganwa River (number 1 in Figure 3-1) and nine sites along three transects in the Nyabugogo natural wetland.

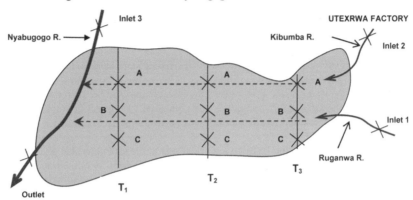

Figure 3-2: Sampling locations, x indicates sampling points and T_x: transect 1, 2 or 3

3.2.3. Sample collection and processing

During each sampling period, three water samples were collected with an interval of two weeks. pH, DO and conductivity were directly measured in the field using WTW pH / Cond / Oxi 340i meter kit. Plants and sediments were collected at only one point per transect because we expected variability to be low due to the short sampling interval (2 weeks). Three samples were taken each period. Sediment samples were collected using a 30 cm nickel-plated carbon steel sampling tube with a core diameter of 1.905 cm. The sediments cores were divided into two sections surface (0 ± 15 cm) and bottom (15 ± 30 cm). The analysis was conducted on the surface section only because we were interested in the part in direct contact with the water. Total heavy metal concentrations (Cd, Cr, Cu, Pb and Zn) were determined in plants, sediments, water, macroinvertebrates and flesh of fish after digestion of the samples, according to the standard analytical procedures. Water samples were digested by mixing 95 ml of water with 5 ml of HNO_3 (65 %) and heated until around 10 ml of the initial solution was obtained. This concentrate was transferred in a clean 100 ml flask; the digestion bottle was then rinsed three times with volumes of 20 ml of distilled water which were added to the concentrate and finally the flask was filled up to the mark with distilled water. Samples of the Papyrus plant (roots, shoots and umbel), flesh of fish (*Clarias* sp. and *Oreochromis* sp.) and Oligochaetes were dried for 24 hours at 103 ($\pm$ 2)°C. A mass of 1.250 g of each dried and grinded sample was digested using a mixture of HNO_3 65 % and H_2O_2 30 %. For sediment samples, a mass of 1.250 g was digested using a mixture of HCl 37 %, HNO_3 65 % and H_2O_2 30 % (Standard Methods, 1992).

The wet acid digestion procedure of heavy metals from solid samples was conducted for at least 7 hours. After digestion the flasks containing samples were left for cooling and then made up to volume by addition of distilled water. All flasks were kept overnight to allow complete settling of particulate matter. The reading on the Flame Atomic Absorption Spectrometer (Perkin Elmer model AAnalyst 200, detection limits: 0.001 mg/L for Cd, 0.008 for Cr, 0.002 for Cu, 0.02 for Pb and 0.004 for Zn) was conducted the following day approximately after 22 hours of settling (Kansiime et al., 2007). Standard heavy metal solutions were prepared according to the specifications of the Atomic Absorption Spectrophotometer (Perkin Elmer, 1996).

3.2.4. Data analysis

Statistical analysis was performed by use of the Excel Stat package. Parametric tests (paired t-test) were used to assess differences in heavy metal concentrations between transects. As the water flows from transect 3 to 1, a comparison between transects was done in the following order: transect 3 was compared to transect 2; transect 2 was compared to transect 1 and finally transect 3 was also compared to transect 1.

Metal accumulation and uptake by living organisms in the swamp is also given by the BioConcentration Factor (BCF). It provides an index of the ability of the organisms to accumulate a particular metal with respect to its concentration in the substrate (Zayed et al., 1998). It was calculated as follows:

$$BCF_i = \frac{[HM]_i}{[HM]_{water}}$$

Where i represents fish, oligochaetes, root, stem or umbel.

3.3. Results

3.3.1. Physico-chemical parameters

Figure 3-3 gives the mean values of in situ measurements of physico-chemical parameters in the water column. The Student t-test did not reveal any significant differences in DO, EC and pH (P > 0.05) between transects. In general, low dissolved oxygen concentrations have been observed in all sampling sites and this pattern persisted throughout all sampling periods. Very low values were recorded in all transects within the swamp, whereas inlet and outlet rivers of the swamp showed slightly higher dissolved oxygen values (Figure 3-3a). A great variation of the conductivity has been recorded in the inlet and outlet rivers of the swamp. High values were recorded in the rivers Kibumba and Ruganwa from the industrial zone whereas low values were recorded in the Nyabugogo River (Figure 3-3b). The pH values remain more or less constant in the wetland; however, higher values were recorded in the Kibumba River (Figure 3-3c).

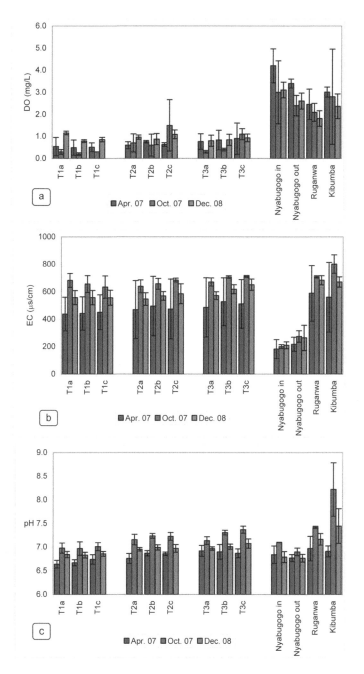

Figure 3-3: Mean values of physico-chemical parameters for different campaigns (n = 3). a) Dissolved oxygen, b) Electrical conductivity and c) pH.

3.3.2. Heavy metals concentrations in the water phase

Cd and Pb concentrations were definitely higher when compared to standards for drinking water (WHO, 2008 and EPA, 2009). However, when comparing with irrigation standards (WHO, 2006), only Cd showed some isolate higher concentrations (Table 3-1). In general no significant differences in heavy metal concentrations between transects were revealed ($P > 0.05$), with the exception of Cr and Pb between transects 1&2 and 1&3 ($P < 0.05$).

Table 3-1: Mean of total metal concentration in water samples (in μg/L) for different campaigns (n = 3); values in bold indicate exceedance of the USEPA (2009) and WHO (2008) drinking water standards; - means below detection limit.

		Cd			Cr			Cu			Pb			Zn		
		Apr 07	Oct 07	Dec 08	Apr 07	Oct 07	Dec 08	Apr 07	Oct 07	Dec 08	Apr 07	Oct 07	Dec 08	Apr 07	Oct 07	Dec 08
WETLAND TRANSECT	T_{1a}	7	3	5	-	13	13	10	11	12	-	-	-	43	26	30
	T_{1b}	7	4	5	-	-	10	12	-	15	-	-	-	36	22	21
	T_{1c}	8	7	6	-	-	-	22	-	10	-	**18**	**27**	18	27	33
	T_{2a}	6	4	4	16	33	19	11	-	13	-	**28**	**28**	15	17	35
	T_{2b}	3	4	7	-	-	19	19	-	17	**31**	-	29	74	11	52
	T_{2c}	-	5	7	-	-	17	14	-	18	**26**	**19**	29	27	116	70
	T_{3a}	-	7	12	-	32	26	14	-	21	-	-	44	43	83	81
	T_{3b}	4	4	8	-	20	22	27	-	22	**22**	-	34	34	4	86
	T_{3c}	18	3	10	-	20	26	115	-	37	**80**	**26**	83	192	5	90
RIVER	Nya ₍in₎	-	2	-	-	17	-	28	-	17	-	-	**15**	107	-	15
	Nya ₍Out₎	1	3	2	-	35	12	26	-	18	-	-	**15**	45	4	26
	Rug	2	5	12	-	-	-	28	-	27	-	-	**23**	55	3	30
	Kib	3	1	10	-	33	34	30	11	34	-	**15**	28	113	12	74
WHO (2006)		10			100			200			5000			2000		
WHO (2008)		3			50			2000			10			3000		
EPA (2009)		5			100			1300			0			5000		

3.3.3. Heavy metals in *Cyperus papyrus* plant

Heavy metal uptake by wetland plants has been shown by many studies. This study focused on *Cyperus papyrus* which is the dominant plant in the Nyabugogo swamp. Analysis of heavy metals contained in the plant showed that *Cyperus papyrus* plays an important role in metal retention. Furthermore, the most important parts of the plant in heavy metal retention were the root system followed by the umbel and finally the stem (Table 3-2). In general no significant differences in Cd, Cr, Cu, Pb and Zn concentrations between transects were observed ($P > 0.05$) with the exception of Cu between transects 1&2 and Pb between transects 1&3. The highest concentrations of Cr, Cu, Pb and Zn found were 45.8 μg Cr / g dry weight, 29.7 μg Cu / g, 61.6 μg Pb / g, 176.8 μg Zn / g and 4.2 μg Cd / g (Table 3-2).

Table 3-2: Mean total metal concentration in Cyperus papyrus (in µg/g DW) for different campaigns (n = 3), values in bold indicate exceedance of the toxic limits according Allen (1989), Amri (2007) and Aksoy et al. (2005); - means below detection limit.

	Cd			Cr			Cu			Pb			Zn		
	Apr 07	Oct 07	Dec 08	Apr 07	Oct 07	Dec 08	Apr 07	Oct 07	Dec 08	Apr 07	Oct 07	Dec 08	Apr 07	Oct 07	Dec 08
T_{1a} Root	**0.8**	**2.4**	**1.4**	**8.8**	**11.2**	**9.8**	0.6	2.2	14.2	**12.0**	**13.8**	**11.3**	88.8	38.6	53.1
T_{1a} Stem	-	-	0.1	**6.4**	**6.0**	**4.0**	-	-	-	0.2	6.2	5.0	56.0	14.8	19.8
T_{1a} Umbel	-	-	-	-	**4.6**	**2.0**	-	-	-	6.0	2.2	2.2	111.4	27.0	23.0
T_{2a} Root	0.2	**2.4**	**2.1**	-	**40.6**	**23.0**	9.4	15.8	**26.0**	**25.2**	**48.6**	**34.8**	78.2	96.4	81.7
T_{2a} Stem	-	0.2	-	-	**6.4**	**4.6**	-	-	4.2	0.4	1.4	1.4	23.8	22.2	37.0
T_{2a} Umbel	-	**0.6**	-	-	**4.8**	**3.1**	-	-	-	0.4	3.8	0.4	38.6	42.8	52.4
T_{3b} Root	**0.8**	**3.4**	**4.2**	-	**45.8**	**36.0**	0.8	**28.6**	**29.7**	**17.8**	**61.6**	**56.1**	81.8	176.8	72.0
T_{3b} Stem	-	-	-	-	**6.2**	**5.3**	-	4.4	3.1	-	0.6	3.7	19.6	11.0	14.6
T_{3b} Umbel	-	**0.6**	-	-	**6.6**	**2.2**	-	4.0	-	0.6	1.6	0.6	27.4	33.8	23.2
Allen (1989)		0.3			n.s.			n.s.			n.s.			n.s.	
Amri (2007)		n.s.			0.3			15			8			n.s.	
Aksoy et al. (2005)		n.s.			n.s.			n.s.			n.s.			200	

n.s.: not specified

3.3.4. Heavy metals in the sediment

Sedimentation has been recognized as the main removal mechanism for heavy metals in natural and constructed wetlands (Walker and Hurl, 2002). However, there are many other processes, including filtration, adsorption, biological assimilation, chemical transformation, and volatilization for mercury as methylated mercury known for its toxicity in the biosphere (Kersten, 1988) that might be involved in reduction of heavy metal concentrations. All heavy metals were found in high concentrations in sediments and especially in transect 3 (Table 3-3), which is the closest to the industrial area. Maximum values were 68.0 µg Cr / g dry weight, 58.3 µg Cu / g, 95.0 µg Pb / g, 244.8 µg Zn / g and 4.2 µg Cd / g. In general, no significant differences in heavy metal concentrations between transects were observed (P > 0.05), with the exception of Cu between transects 2&3 and Pb between transects 1&2.

Table 3-3: Mean total heavy metal concentration in sediment samples (in µg/g DW) for different campaigns (n = 3); values in bold indicate exceedance of the standards according MacDonald et al. (2000) and Smolders et al. (2003)

	Cd			Cr			Cu			Pb			Zn		
	Apr 07	Oct 07	Dec 08	Apr 07	Oct 07	Dec 08	Apr 07	Oct 07	Dec 08	Apr 07	Oct 07	Dec 08	Apr 07	Oct 07	Dec 08
T₁ₐ S	1.2	4.2	**2.9**	**22.4**	**54.2**	**48.0**	18.8	**40.2**	**41.1**	21.8	**56.0**	**58.6**	58.8	71.2	87.2
T₂ₐ S	0.8	**3.0**	**4.1**	19.2	**35.0**	**52.0**	29.8	**32.8**	**37.0**	26.0	**37.4**	**42.1**	50.6	72.2	65.2
T₃ₐ S	1.2	**3.0**	**3.9**	9.8	**59.2**	**68.0**	46.6	**52.4**	**58.3**	74.4	**95.0**	**74.9**	**188.0**	**244.8**	**124.0**
MacDonald et al (2000)		0.99			43.4			31.6			35.8			121	
Smolders et al (2003)		n.s.			n.s.			n.s.			n.s.			200	

n.s.: not specified

3.3.5. Heavy metals in fish and invertebrates

As argued by Van Straalen et al. (2005), the concentration of a pollutant inside an organism is a good indicator of its bioavailability. Therefore, heavy metal analysis was carried out on living organisms as indicated in the material and methods section. Except Cu and Zn, all other heavy metals were found in high concentrations in *Clarias* sp., *Oreochromis* sp. and Oligochaetes. Results are presented in Table 3-4.

Table 3-4: Mean total metal concentration in fish (Clarias sp. and Oreochromis sp.) and macroinvertebrates (Oligochaetes) in µg/g dry weight (n = 3). Values in bold indicate exceedance of the standards according Denny et al. (1995), EC (2001) and Maret et al. (2003)

	Cd	Cr	Cu	Pb	Zn
Clarias sp.	**2.85**	**6.35**	2.87	**0.67**	5.11
Oreochromis sp.	**2.78**	**5.12**	1.82	**0.81**	5.52
Oligochaetes*	0.33	**22.11**	3.21	1.76	9.06
Denny et al. (1995)	-	0.01 – 1.0	0.1 – 50	-	2 – 100
EC (2001)	-	1 – 10*	20 – 200*	-	40 – 500*
Maret et al. (2003)	0.05	-	-	0.2	-
	0.08 - 2.32*	-	-	0.2 - 7.09*	-

3.4. Discussion

3.4.1. Heavy metal concentrations in the water phase

From this study, there seems in general a pollution problem caused by Cd and Pb with high concentrations when compared to drinking water and irrigation standards (Table 3-1). Concentrations of Cr, Cu and Zn were below the EPA (2009) drinking water contaminant limits, WHO (2008) guidelines for drinking water quality and WHO (2006) guidelines for the safe use of wastewater, excreta and grey water. Both Cd and Pb showed high concentrations in transect 3. This can be explained by the intensive activities of car garages present there. Cd is known to be the major inorganic used in paint industry. It could thus be expected that car painting activities may be the major source of Cd pollution recorded at that location of the swamp. The Pb pollution could be traced to the garages wastewater discharge containing oil directly dumped in the swamp. Furthermore, physico-chemical parameters like pH, redox potential and salinity affect the (im)mobility of heavy metal as reported by Kelderman et al. (2000), Kelderman and Osman (2007) and Du Laing et al. (2008). Analysis conducted on the samples collected from the swamp showed in general a high salinity with mean values between 27.5 – 102.5 mg/L of Cl⁻ (data not shown). This could also explain the high Cd concentrations found in the water. In general no seasonal differences were observed as one could expect to have lower metal concentrations during the rainy season due to the dilution effect of rain waters.

Table 3-5: Reported heavy metal concentrations in water from different wetlands

Sites	pH	EC (µS/cm)	Heavy metals (µg/L)				Reference
			Cd	Cu	Pb	Zn	
Kahendero swamp (Uganda)	6.90	2600.0	n.s.	n.s.	n.s.	n.s.	Denny et al., 1995
Hamukungu bay (Uganda)	9.80	253.0	n.s.	189.0	n.s.	60.0	"
Lake George (Uganda)	9.90	215.0	n.s.	11.2	n.s.	n.s.	"
Mweya (Uganda)	9.60	210.0	n.s.	22.0	n.s.	n.s.	"
Hamukungu bay (Uganda)	n.s.	n.s.	n.s.	1.0	n.s.	n.s.	Monday et al., 2003
Mweya (Uganda)	n.s.	n.s.	n.s.	n.s.	n.s.	n.s.	"
Bushatu (Uganda)	n.s.	n.s.	n.s.	1.0	n.s.	6.0	"
Lake Edward (Uganda)	n.s.	n.s.	n.s.	1.0	n.s.	1.0	"
Bwaise (Uganda)	6.78	878.0	n.s.	25.0	50.0	220.0	Nabulo et al., 2008
Kinawataka (Uganda)	5.85	187.3	n.s.	40.0	n.s.	160.0	"
Namuwongo (Uganda)	7.15	566.7	n.s.	50.0	20.0	160.0	"
Murchison (Uganda)	8.86	585.0	n.s.	50.0	n.s.	50.0	"
Nyabugogo River (Rwanda)	6.82	182.0	1.0	28.0	n.s.	106.5	This study
Nyabugogoswamp (Rwanda)	7.02	595.0	7.5	20.4	28.3	53.7	"
Kibumba River (Rwanda)	8.22	802.0	1.0	10.5	15.0	11.8	"

n.s. = not specified

Table 3-5 gives an overview of some reported heavy metal concentrations in water from different wetlands in the region. These values are given for comparison with the pollution in the Nyabugogo wetland. Heavy metal pollution depends highly on the socio-economic development surrounding the wetland and its physico-chemical conditions. Therefore, the reported heavy metal concentrations can be taken as values recorded under specific local conditions. These may vary site to site and period to period. Results of this study on the Nyabugogo swamp showed in general lower concentrations when compared to other sites monitored in the region.

3.4.2. Heavy metals in *Cyperus papyrus* plant

From this study, analysis of *Cyperus papyrus* showed that metals accumulated preferably on the roots in the following order Cu > Zn > Pb > Cr > Cd. However, uptake to the stem and translocation to the umbel was also noticed (Table 3-2). Our findings on heavy metal accumulation by *Cyperus papyrus* are in good agreement with what has been reported by other researchers (Mays et al., 2001; Nabulo et al., 2008). There is a pollution problem by Cd and Cr in all transects within the swamp. Aksoy et al. (2005) reported that cadmium is taken up metabolically and easily transported to other parts of the plant. In this study, Cd was hardly taken up and translocated in the plant whereas Cr was easily taken up and translocated. The difference observed between Cd and Cr accumulation confirms uptake of heavy metals depends significantly on the metal as well as water conditions such as salinity, pH, redox potential and organic matter content (Kelderman et al., 2000; Kelderman and Osman, 2007; Du Laing et al., 2008). Furthermore, even if Cr and Pb are not essential for plant growth, some studies have indicated that at low concentrations (1μM), Cr stimulates plant growth (Bonet et al., 1991). Regarding lead, it can move through root tissue, mainly via the apoplast and radially through the cortex where it accumulates near the endoderm. The endoderm acts as a partial barrier to the translocation of Pb through the root to the shoot. This may be one of the reasons for the much greater accumulation of Pb in roots than in shoots (Jones et al., 1973; Verma and Dubey, 2003).

According to Manios et al. (2003), several authors have shown that there is a threshold of tolerance of each plant to heavy metals accumulation. For a number of environmentally, physiologically and genetically determined reasons this threshold is different among plant species. When this limit is passed, the toxic effect of metals in plants takes place and metals become poisonous. Concentration limits for Cd and Cr are 0.3 μg / g dry weight in plants in unpolluted environments (Allen, 1989; Amri et al., 2007). Results of this study are above that limit, showing that the Nyabugogo swamp is a metal polluted environment. Comparison between heavy metals accumulated by roots indicates clearly that there is a decreasing pattern in the heavy metal concentration from transect 3 towards transect 1 in the swamp (Table 3-2).

Concentrations of Cu, Pb and Zn were in general below toxic limits (Table 3-2). However, some isolate high concentrations were recorded for Cu and Zn. Cyperus papyrus accumulated, took up and translocated more Zn when compared to the other heavy metal. This is explained by the fact that Zn like Cu and Fe is an essential micronutrient for plant metabolism; but when present in excess, it becomes extremely toxic (William et al., 2003). Aksoy et al. (2005) and Amri et al. (2007) gave a limit of 100 μg / g dry weight above which Zn becomes poisonous to the plant. Furthermore, this study found that Cu like Cd was strongly accumulated on the root. Very few quantities were taken up and translocated to the upper part of the plant. The limit for Pb is 8 μg / g dry weight (Amri, 2007). According to Roos (1994), a limit between 30 and 100 μg / g dry weight of Pb is considered to be toxic to the plants. In transect 3, these values were not exceeded; but the results obtained are one to eight times Amri's limit. The highest value was found in transect 3. The Cu limit is 15 μg / g dry weight (Amri et al., 2007). Above this limit, Cu is regarded as poisonous to the plants. Considering values found in Papyrus (Table 3-2), high values of Cu were recorded in transects 2 and 3.

3.4.3. Heavy metals in the sediment

This study showed a contamination of the sediment by Cd, Cr, Cu, Pb and Zn especially in transect 3 (Table 3-3). In general the accumulation of metals in the sediment did not depend on the season, though one could expect to have higher accumulation in the dry season rather than in the rainy season due to the flush out occurring in the swamp in that period. Based on the threshold effect concentration (TEC) reported by MacDonald et al. (2000) (Table 3-3); which gives the concentration limits below which harmful effects are unlikely to be observed, it is clear that all heavy metals studied are above the given limits. Therefore, the Nyabugogo sediment can be classified as polluted by heavy metals. Moreover, it is a potential source of metal contamination back into the water, as sediments contain a pool of pollutants that can be mobilized or immobilized if certain physico-chemical changes occur (Kelderman et al., 2000; Kelderman and Osman, 2007).

3.4.4. Heavy metals in fish and invertebrates

The mean concentrations of metals in the flesh of fish were in general above the limits. This study shows that the Cr content in *Clarias* sp., *Oreochromis* sp. and Oligochaetes is high and presents a risk (Denny et al., 1995). Cd and Pb values are also high in fish when compared to the limits fixed by the European Commission on heavy metals in fish (2001). Concerning Oligochaetes, Cd and Pb concentrations are not beyond the ranges given by Maret et al. (2003) (Table 4). Food ingestion is normally a more important source of metals contamination than drinking water. Therefore, the accumulation of metals in tissue can result from fish eating habits (Monday et al., 2003). Oligochaetes live in sediments and feed on it. As it is known, fish like *Oreochromis* sp. feed also on sediment containing Oligochaetes and *Clarias* sp. is a carnivorous species and feeds on Oreochromis sp. It is clear how the ingestion of metal contaminated Oligochaetes and the direct contact with contaminated water might be the principal route of heavy metals exposure to fish in wetlands.

Table 6 gives reported heavy metal concentrations in fish in the region compared to the Northeast Mediterranean sea. Results of this study showed higher Cd and Cr concentrations in fish. However, Pb concentration was lower when compared to the reported concentrations found by Kaley et al. (1999). Even if the Zn concentration was lower in this study when compared to the reported values of Dennis et al. (1995), Kaley et al. (1999) and Monday et al. (2003), heavy metal concentrations from the present study are higher than the standard limit as discussed earlier and thus present a human health concern for people consuming fish from the swamp.

Table 3-6: Reported heavy metal concentration in fish.

Sites	Fish species	Heavy metals (μg/g)					Reference
		Cd	Cr	Cu	Pb	Zn	
Hamukunga bay (Uganda)	*Oreochromis leucostictus*	n.s.	n.s.	549.6	n.s.	49.4	Denny et al. 1995
Mweya (Uganda)	*Oreochromis leucostictus*	n.s.	n.s.	117.0	n.s.	16.6	"
	Oreochromis niloticus	n.s.	n.s.	16.3	n.s.	44.5	"
Northeast Mediterranean	*Caranx crysos*	1.36	2.07	6.15	7.50	33.6	Kaley et al. 1999
sea (Turkey)	*Mugil cephalus*	1.07	1.35	5.12	7.33	30.9	"
	Mullus barbatus	1.43	1.91	5.88	9.11	25.8	"
Kazinga Edward (Uganda)	*Clarias* sp.	n.s.	n.s.	0.6	n.s.	14.0	Monday et al. 2003
	Oreochromis leucostictus	n.s.	n.s.	n.s.	n.s.	n.s.	"
	Oreochromis niloticus	n.s.	n.s.	0.6	n.s.	8.3	"
North George (Uganda)	*Clarias* sp.	n.s.	n.s.	2.5	n.s.	16.3	"
	Oreochromis leucostictus	n.s.	n.s.	0.2	n.s.	12.9	"
	Oreochromis niloticus	n.s.	n.s.	0.8	n.s.	15.2	"
South George (Uganda)	*Clarias* sp.	n.s.	n.s.	0.7	n.s.	23.9	"
	Oreochromis leucostictus	n.s.	n.s.	0.5	n.s.	22.5	"
	Oreochromis niloticus	n.s.	n.s.	0.5	n.s.	11.3	"
Nyabugogo swamp	*Clarias* sp.	2.85	6.35	2.87	0.67	5.11	This work
(Rwanda)	*Oreochromis niloticus*	2.78	5.12	1.82	0.81	5.52	"
	Oligochaetes	0.33	22.11	3.21	1.76	9.06	"

n.s. = not specified

3.4.5. Bioconcentration factor

In this study bioconcentration factors show that heavy metals are generally accumulated more in plants than in fish and Oligochaetes (Figure 3-4).

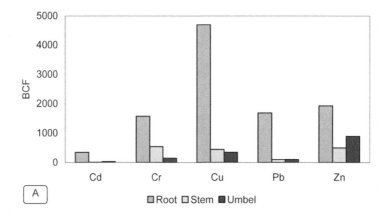

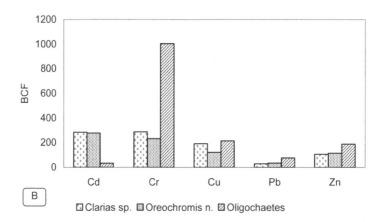

Figure 3-4: Bioconcentration factors of heavy metals in different parts of Cyperus papyrus (A) and in Clarias, Oreochromis and Oligochaetes (B).

Oligochaetes accumulated more Cr, Cu, Pb and Zn when compared to fish which accumulated more Cd. This could be explained by the fact that Cr, Cu, Pb and Zn were in higher concentrations in the sediment when compared to Cd (Table 3-3). Fish accumulated more Cd probably due to the higher Cd concentration found in the water phase (Table 3-1). Based on the comparison between the BCF of plants, fish and macroinvertebrates, one could suggest the use of *Cyperus papyrus* as bioindicator species for metal pollution in wetlands.

3.4.6. Physico-chemical parameters

This study showed that pH values recorded in the Ruganwa and Kibumba rivers were higher than 7. However, pH values recorded within the swamp and in the Nyabugogo River were more or less constant close to 7. In general, the pH of Rwandan rivers varies between pH 5 – 6. This is due to the geology of the Rwandan soil (Rutunga et al., 1998). The highest pH values found in the Kibumba and Ruganwa rivers which flow from the textile factory and industrial park of Gikondo are explained by the use of alkaline reagents by factories in their processes. While comparing pH values in different transects, a decrease was recorded from transect 3 towards transect 1. The same pattern was observed for dissolved oxygen (DO). However, the electro-conductivity did not vary as much as the DO or pH. The decrease in pH might be explained by the presence of dead plant material trapped in the dense mat of papyrus covering the surface of the swamp. This is playing an important role in retaining dead material. Others researchers (Verhoeven, 1986; Azza et al., 2000) have also noticed such trends in the pH variation in a wetland.

The decomposition of dead material increases the organic content within the wetland. This will play an important role in the consumption of the dissolved oxygen within the wetland and explains the very low values recorded from transect 3 to transect 1. The organic matter plays a major role in adsorbing and complexing some cations present in the water phase, thus resulting in a decrease of the electro-conductivity of the water. As reported by Rogers et al. (1991); Kansiime and Nalubega (1999), the structure and permeability of the roots system of the floating mat play an important role in the degradation of sewage flowing underneath the mats. Direct contact of polluted water with the roots is important for pollutant removal from wastewater.

3.5. Conclusions

The metal concentration in water flowing through the Nyabugogo swamp has been determined in water, in *Cyperus papyrus*, in sediment, in fish (*Clarias* sp. and *Oreochromis* sp.) and in Oligochaetes. Although the results showed that there is a considerable metal accumulation in plants and in sediments, the study showed that there is also a high concentration of heavy metals in the water body of the swamp and in the outflow. There is in general a pollution problem caused by Cd and Pb in the water phase. The highest accumulation of heavy metals has been found both in the plants and the sediments. Roots play a more dominant role in heavy metals removal compared to the stem and the umbel. Fish in the Nyabugogo swamp showed a high concentration of heavy metals especially Cr, Cd and Pb. In general, there is a genuine concern for human health through direct consumption and uptake of metals via drinking water and eating fish, but also indirectly by using the polluted water for irrigation of crops around the wetland.

Acknowledgements

The authors are grateful to the NUFFIC / Dutch Government and the National University of Rwanda for the financial support provided for this research through the Netherlands Programme for Institutional Strengthening of Post-secondary Education and Training Capacity (NPT / RWA / 051-WREM) Project. The authors would like also to thank the laboratory staff for UNESCO-IHE for their technical assistance during this research.

3.6. References

Aksoy, A., Demirezen, D. & Duman, F. (2005). Bioaccumulation, detection and analyses of heavy metal pollution in Sultan Marsh and its environment. Water, Air and Soil Pollution, 164, 241 – 255.

Allen, S.E. (1989). Chemical Analysis of Ecological Materials, 2nd ed. Blackwell Scientific Publications. Oxford.

American Public Health Organisation, (1992). Standards Methods for examination of Water and Wastewater 18th edition.

Amri, N., Benslimane, M., Zaoui, H., Hamedoun, M. & Outiti, B. (2007). Evaluation Of The Heavy Metals Accumulate In Samples Of The Sediments, Soils And Plants   By   ICPOES   With   The   Average   Sebou. http://www.fsr.ac.ma/MJCM/vol8-art08.pdf (Assessed 10 August 2009).

Blackwell, M.S.A., Hogan, D.V. & Maltby, E. (2002). Wetlands as regulators of pollutant transport. In: Haygarth, PM., Jarvis, SC. (Eds.), Agriculture, Hydrology and Water Quality. CAB International.

Bonet, A., Poschenrieder, C. & Barcelo, J. (1991). Chromium III - Iron interaction in Fe - deficient and Fe - sufficient bean plants. I. Growth and nutrient content. Journal of Plant Nutrition, 14, 403 – 414.

Deng, H., Ye, Z.H. & Wong, M.H. (2004). Accumulation of lead, zinc, copper and cadmium by 12 wetland plant species thriving in metal-contaminated sites in China. Environmental Pollution, 132, 29 – 40.

Denny, P. (1995). Heavy metal contamination of Lake George (Uganda) and its wetlands. Hydrobiologia, 257, 229 – 239.

Denny, P. (1987). Mineral cycling by wetland plants – a review. Arch. Hydrobiol. Beich. Ergebn. Limnol. 27, 1 – 25.

Du Laing, G., De Vos, R., Vandecasteele, B., Lesage, E., Tack, F.M.G. & Verloo, M.G. (2008). Effect of salinity on heavy metal mobility and availability in intertidal sediments of the Scheldt estuary. Estuarine, Coastal and Shelf Science, 77, 589 – 602.

EPA (2009). Online series. Drinking Water contaminants. Specific Fact Sheets for Consumer. http://www.epa.gov/safewater/contaminants/index.html. (Assessed 5 November 2009).

EC (2001) Commission Regulation as regards heavy metals, Directive 2001/22/EC, No: 466/2001.

Ewers, U. & Schlipkoter, H.W. (1991). Intake, distribution and excretion of metal compound in humans and animals. In: Merian E. Metals and Their Compound in the environment; Occurrence, Analysis and Biological Relevance. VCH Weiheim: 571 – 583.

Fleming, C.A. & Trevors, J.T. (1989). Copper toxicity and chemistry in the environment: A Review. Water, Air and Soil Pollution, 44, 143 – 158.

Fritioff, A., Kautsky, L. & Greger, M. (2005). Influence of temperature and salinity on heavy metal uptake by submersed plants. Environmental Pollution, 133, 265 – 274.

Jones, L.H.P., Clement, C.R. & Hopper, M.J. (1973). Lead uptake from solution by perennial ryegrass and its transport from roots to shoots. Plant Soil, 38, 403 – 414.

Judith, S.W. & Peddrick, W. (2004). Metal uptake, transport and release by wetland plants: implications for phytoremediation and restoration. Environment International, 30, 685 – 700.

Kalay, M., Ay, Ö. & Canli, M. (1999). Heavy Metal Concentrations in Fish Tissues from the Northeast Mediterranean Sea. Bulletin of Environmental Contamination and Toxicology, 63, 673 – 681.

Kansiime, F. & Nalubega, M. (1999). Wastewater Treatment by a Natural Wetland: the Nakivubo Swamp, Uganda, Processes and Implications, A.A. Balkema, Rotterdam.

Kansiime, F., Saunders, M.J. & Loiselle, S.A. (2007). Functioning and dynamics of wetland vegetation of Lake Victoria: an overview. Wetlands Ecology and Management, 15, 443 – 451.

Kelderman, P., Drossaert, W.M.E., Zhang, M., Galione, L., Okwonko, C. & Clarisse, I.A. (2000). Pollution assessment of the canal sediments in the city of Delft (The Netherlands). Water Research, 34, 936 – 944.

Kelderman, P. & Osman, A.A. (2007). Effect of redox potential on heavy metal binding forms in polluted canal sediments in Delft (The Netherlands). Water Research, 41, 4251 – 4261.

Kersten, M. (1988). Geochemistry of priority pollutants in anoxic sludges: Cadmium, arsenic, methyl mercury, and chlorinated organics, in Salomons, W., and Forstner, U., eds., Chemistry and biology of solid waste: Berlin, Springer-Verlag, 170 – 213.

MacDonald, D.D., Ingersoll, C.G. & Berger, T.A. (2000). Development and evaluation of consesus-based sediment quality guidelines for freshwater ecosystems, Arch. Contam. Toxicol., 39, 20 – 31.

Manios, T., Stentiford, E.I. & Millner, P.A. (2003). The effect of heavy metals accumulation on the chlorophyll concentration of *Typha latifolia* plants, growing in a substrate containing sewage compost and watered with metaliferous water, Ecological Engineering, 20, 65 – 74.

Maret, T.R., Cain, D.J., Mac Coy, D.E. & Short, T.M. (2003). Response of benthic invertebrate assemblages to metal exposure and bioaccumulation associated with hard-rock mining in northwestern streams, USA. Journal of the North American Benthological Society, 22 (4), 598 – 620.

Mays, P.A. & Edwards, G.S. (2001). Comparison of heavy metal accumulation in a natural wetland and constructed wetlands receiving acid mine drainage. Ecological Engineering, 16, 487 – 500.

Mitsch, W.J. & Jørgensen, S.E. (1989). Ecological Engineering: An Introduction to Ecotechnology. John Wiley & Sons, New York.

Mitsch, W.J. & Jørgensen, S.E. (2004). Ecological Engineering and Ecosystem Restoration. John Wiley & Sons, Inc., New York.

Monday, S.L., Kansiime, F., Denny, P. & James, S. (2003). Heavy metals in Lake George, Uganda, with relation to metal concentrations in tissues of common fish species. Hydrobiologia, 499, 83 – 93.

Nabulo, G., Oryem O.H., Nasinyama, G.W. & Cole, D. (2008). Assessment of Zn, Cu, Pb and Ni contamination in wetland soils and plants in the Lake Victoria basin. International Journal of Environmental Science and Technology, 5 (1), 65 – 74.
http://ijest.indexcopernicus.com/abstracted.php?level=4&idissue=751301 (Assessed 18 January 2010)

Perkin Elmer (1996). Analytical Methods for Atomic Absorption Spectroscopy.

Rutunga, V., Kurt, G.S., Karanja, N.K., Gachene, C.K.K. & Nzabonihankuye, G. (1998). Continuous fertilization on non-humiferous acid Oxisols in Rwanda "Plateau Central": Soil chemical changes and plant production. Biotechnol. Agron. Soc. Environ., 2 (2), 135 – 142.

Smolders, E., Waegeneers, N., Lison, D., Veroughstraete, V. & De Backer, L. (2003). The EC risk assessment of Cadmium (report available on http://www.icsu-scope.org/cdmeeting/2003meeting/abs_Smolders_etal.htm) (Assessed 21 October 2009).

Van Dam, A.A., Dardona, A., Kelderman, P. & Kansiime, F. (2007). A simulation model for nitrogen retention in a papyrus wetland near Lake Victoria, Uganda (East Africa). Wetlands Ecology and Management, 15, 469 – 480.

Van Straalen, N.M., Donker, M.H., Vijver, M.G. & Van Gestel, C.A.M. (2005). Bioavailability of contaminants estimated from uptake rates into soil invertebrates. Environmental Pollution, 136, 409 – 417.

Verhoeven, J.T.A., Arheimer, B., Yin, C.Q. & Hefting, M.M. (2006). Regional and global concerns over wetlands and water quality. Trends Ecol. Evol., 21, 96 – 103.

Verma S., & Dubey R.S. (2003). Lead toxicity induces lipid peroxidation and alters the activities of antioxidant enzymes in growing rice plants. Plant Science, 164, 645 – 655.

Walker, J.D. & Hurl, S. (2002). The reduction of heavy metals in stormwater wetland, Ecological Engineering, 18, 407– 414.

Williams, L.E., Pittman, J.K. & Hall, J.L. (2000). Emerging mechanisms for heavy metal transport in plants. Biochimca et Biophysica Acta, 1465 (12), 104 – 26.

WHO (2006). Guidelines for the safe use of wastewater, excreta and grey water. Volume 2 wastewater use in agriculture.

WHO (2008). Guidelines for Drinking-Water Quality, 3rd edition First Addendum to the 3rd edition Volume 1 recommendations (Vol. 1) WHO, Geneva, Recommendations.

Zayed, A., Gowthaman, S. & Terry, N. (1998). Phytoaccumulation of trace elements by wetland plants: I. Duckweed. Journal of Environmental Quality, 27, 715 – 721.

Zedler, J.B. & Kercher, S. (2005). Wetland resources: status, trends, ecosystem services, and restorability. Annual Review of Environmetal Resources, 30, 39 – 74.

Chapter 4: Heavy metal removal in duckweed and algae ponds as a polishing step for textile wastewater treatment

Main parts of this chapter have been published as:

Rousseau, D.P.L., **Sekomo, C.B.,** Saleh, S.A.A.E., & Lens, P.N.L., (2011) Duckweed and Algae Ponds as a Post-Treatment for Metal Removal from Textile Wastewater. In: Vymazal J. (Ed), Water and Nutrient Management in Natural and Constructed Wetlands, Springer Science, Dordrecht.

Sekomo, C.B., Rousseau, D.P.L., Saleh, S.A.A.E., & Lens, P.N.L., (2012b). Heavy metal removal in duckweed and algae ponds as a polishing step for textile wastewater treatment. Ecological Engineering 44, 102 – 110.

Abstract

Untreated textile wastewater is a typical source of heavy metal pollution in aquatic ecosystems. In this study, the use of algae and duckweed ponds as post-treatment for textile wastewater has been evaluated under the hypothesis that differing conditions such as pH, redox potential and dissolved oxygen in these ponds would lead to different heavy metal removal efficiencies. Two lab-scale systems each consisting of three ponds in series and seeded with algae (natural colonization) and duckweed (*Lemna minor*), respectively, have been operated at a hydraulic retention time of 7 days and under two different metal loading rates and light regimes (16/8 hours light/darkness and 24 hours light). Cr removal rates were 94 % for the duckweed ponds and 98 % for the algal ponds, indifferently of the metal loading rate and light regime. No effect of pond type could be demonstrated for Zn removal. Under the 16/8 light regime, Zn removal proceeded well ($\sim$ 70 %) at a low metal loading rate, but dropped to below 40 % at the higher metal loading rate. The removal efficiency raised back to 80 % at the higher metal loading rate but under 24 hours light regime. Pb, Cd and Cu all showed relatively similar patterns with removal efficiencies of 36% and 33% for Pb, 33% and 21% for Cd and 27% and 29% for Cu in the duckweed and the algal ponds, respectively. This indicates that both treatment systems are not very suitable as a polishing step for removing these heavy metals. Despite the significant differences in terms of physico-chemical conditions, differences in metal removal efficiency between algal and duckweed ponds were rather small.

Key words: Algae, duckweed, textile wastewater, heavy metals, tracer test

4.1. Introduction

Heavy metal contamination of the environment has increased sharply since 1900 (Nriagu, 1979) and actually the increasing contamination of freshwater systems with thousands of industrial and natural chemical compounds is one of the major environmental and human health problems worldwide (Ensley, 2000; Schwarzenbach et al., 2006). According to Johnson and Hallberg (2005) for example, it was estimated that in 1989 ca. 19,300 km of streams and rivers, and ca. 72,000 ha of lakes and reservoirs worldwide had been seriously damaged by mine effluents, although the true scales of the environmental pollution caused by mine water discharge is difficult to assess accurately. Other well-known anthropogenic sources of heavy metal pollutants are smelting of metalliferous ores, electroplating, gas exhaust, energy and fuel production, the application of fertilizers and municipal sludge to land, and industrial manufacturing as the textile industry (Raskin et al., 1994; Cunningham et al., 1997; Blaylock and Huang, 2000).

Textile wastewater is a mixture of colorants (dyes and pigments) and various organic compounds used as cleaning solvents, and has a high chemical as well as biological oxygen demand. It also contains high concentrations of heavy metals and total dissolved solids (Sharma et al., 2007). For example, its discharge led to the complete disappearance of submerged and free floating hydrophytes, as well as also to the disappearance of certain marshy species in the pools in the area of Sanganer town, Jaipur.

Several technologies are available to remove heavy metals (HM) from wastewater such as chemical precipitation, flotation, coagulation-flocculation, ion exchange and membrane filtration (Kurniawan et al., 2006); all have their advantages and limitation in application, require high capital investment as well as tend to generate a sludge disposal problem (Cohen, 2006; Aziz et al., 2008). Thus for most developing countries, alternative technologies are needed that are within the economical and technological capability of these nations.

Wetland systems have been used as an alternative cost-effective technology to conventional wastewater treatment methods. It has been shown that they can efficiently remove heavy metals from both domestic and industrial wastewater (Rodgers and Dunn, 1992; Tang, 1993; Lakatos et al., 1997; Qian et al., 1999 and Le Duc and Terry, 2005). Plants play important roles in these systems for the removal of pollutants (Brix, 1994, Oporto et al., 2006). They not only take up nutrients, but also host micro-organisms which in turn provide sites for metal sorption as well as carbon sources for bacterial metabolism (Jacob and Otte, 2004; Marchand et al., 2010). Water hyacinth (*Eichhornia crassipes*) and duckweed (*Lemna sp.*) are commonly used in aquatic treatment systems. These plants and algae influence the redox and pH conditions of the aquatic systems as a result of photosynthesis and respiration processes (Shilton, 2005). Therefore, it is expected that they will also contribute to the metal removal processes. Given that pH and redox fluctuations are much higher in algal ponds than in duckweed ponds (Vymazal, 1995 and Kadlec and Wallace, 2008), one may expect different metal removal efficiencies in these pond systems.

In wetlands, the pH is among the parameters playing a key role in metal removal processes. The pH of the solution results in the oxidation or reduction of metallic species, therefore metal speciation, solubility, transport, and eventual bioavailability of metals in aqueous solutions is influenced within the wetland (Sheoran and Sheoran, 2006). The solubility of metal hydroxide minerals for instance increases with decreasing pH, and more dissolved metals become potentially available for incorporation in biological processes as pH decreases. Galun et al. (1987) also reported that the pH affects the solution chemistry of the metals, the activity of the functional groups in the biomass and the competition of metallic ions. This has in turn a direct effect on the adsorption, desorption, precipitation and co-precipitation process of metallic ions on the biosorbent.

Redox conditions affect the mobility of metals in two ways. Firstly, there are direct changes in the oxidation states of certain metals. Secondly, redox conditions may indirectly influence the mobility of metals that occur in only one oxidation state (e.g. cadmium and zinc). This is the case when these metals are bound to elements (sulphides, organic matter and some chelating agents) that are labile (Stigliani, 1992). For heavy metals in wastewater, the effect mainly occurs to the heavy metals bound to reducible or oxidisable compounds.

In this study, we are presenting results of one step from a study that aspires to devise a sustainable low-cost heavy metals treatment technology for textile wastewater based on a two-step process: (1) heavy metals precipitation after sulphate reduction in anaerobic bioreactors and (2) a polishing step by means of aquatic treatment systems. The whole system constitutes an integrated system for heavy metals treatment (Sekomo et al., 2011b). The present study focuses on the second step, with the objective to study the effect of heavy metal loads on their removal efficiency and to find out how this is influenced by the light regime, pH and redox potential conditions in duckweed (DP) and algae ponds (AP).

4.2. Materials and methods

4.2.1. Laboratory set-up

The set-up comprised two treatment lines, one with AP and one with DP, each consisting of three glass aquaria in series (L×W×D: 50×30×30cm; maximum water depth 23.5cm) as described by Shi et al. (2010). They were coded A1 – A3 for the AP, and D1 – D3 for the DP (Figure 4-1).

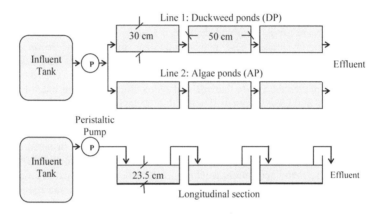

Figure 4-1. Schematic representation of the AP and DP set-up.

The hydraulic behaviour of the ponds was studied using a tracer continuous loading technique using lithium chloride (LiCl) (Zimmo et al., 2000). Initial background of the tracer concentration in the pond water was measured in all sampling points immediately prior to the tracer addition. The flow pattern was characterised according the models described by Levenspiel (1972) and García et al. (2004).

4.2.2. Wastewater characteristics

Synthetic wastewater was pumped from a holding tank at equal rates to the AP and DP. This wastewater was based on the Hunter nutrient solution (Leman, 2000), whose general characteristics are: 17.7 mg/L of TOC, 4.6 mg/L of NH_4^+-N, 4.8 mg/L NO_3^- and 7.3 mg/L of PO_4^{3-} and spiked with Merck heavy metal standard solutions (1000 mg/L) to a level equal to what can be expected of anaerobically pre-treated textile wastewater (see 2.4; based on Sekomo et al., 2011b). The hydraulic retention time (HRT) was set to seven days; evapotranspiration losses were compensated daily by adding demineralised water.

4.2.3. Operating conditions

The system was initially operated under 16 / 8 hours light / dark regime with a light intensity of 125 μE/m^2.s on the water surface. The DP were inoculated with *Lemna minor* species at a density of 600 g fresh weight/m^2 collected in Delft canal water. Analysis conducted on sediments of these canals showed these are a heavy metal polluted site (Kelderman et al., 2000). AP were inoculated with Delft canal water at start-up for algae (natural colonisation) development. After an initial start-up period of two weeks to allow plant and algae growth, three experiments were done: (1) Cd, Cr, Cu, Pb, and Zn at influent concentrations of 0.05, 1.5, 0.1, 0.25 and 1.25 mg/L respectively, named Run 1; (2) double metal concentrations of Run 1, named Run 2; and (3) double metal concentrations of Run 1 and switch to a 24 hours light regime, named Run 3. All experiments were done for 3 weeks (i.e. 3 times the HRT) to allow the system to reach steady state conditions. Duckweed biomass was harvested every five days to restore the duckweed density to 600 g fresh weight/m^2. This density was selected to prevent overcrowding and to maintain sufficient cover to minimize the development of algae in duckweed ponds (Zimmo, 2003). Floating algae formed over the water surface were collected regularly to improve oxygen levels in the system since the thick layer was obstructing light penetration.

4.2.4. Sampling, preservation and analytical procedures

Dissolved oxygen (DO), pH, was measured using HACH HQ10–LDO (Luminescent Dissolved Oxygen) and pH, temperature and ORP measurements were done with a WTW pH 340i. All equipments were properly calibrated following the manufacturer instructions. Readings were taken at 10 cm below the water surface every hour (one pond / 24 hours) during the experimental periods to produce physicochemical profiles for each pond under the different conditions of each run. As the system was operated under constants conditions of flow and temperature, several samples (N=3) were taken to determine the degree of variations in the system.

Sampling, preservation and analytical procedures followed the Standard Methods for Water and Wastewater Examination (APHA, 1995). Once a week, after good mixing, 250 ml of composite water samples were taken from tanks collecting all effluent during that week for mass balance purposes. Grab samples were collected from the influent and the effluents of each pond at 4 instances during the last two weeks of each run. During the first 10 days of each run, no grab samples were taken as it was assumed that the system was adapting to the newly applied influent conditions.

Biofilm accumulated on the walls of the ponds, floating algae and duckweed biomass were harvested and stored in a freezer until further HM analysis. Those samples were first dried at 105 °C, and then the biomass was digested with nitric acid using a microwave technique. HM analysis (Cd, Cr, Cu, Pb and Zn) was conducted for all water samples by using the flame and flameless atomic absorption spectrometer models Perkin-Elmer 3110 and Thermo elemental – SOLAR 95 / Furnace.

4.2.5. Data treatment

Data were treated with Microsoft Excel^{TM} and Excel Stat pro software. Data were tested for normality and then the t-test and ANOVA were conducted on them. Where the normality test wa not succefull, the Kruskal-Wallis non-parametric test was conducted. For physico-chemical data, the t-test ($\alpha = 0.05$) and for heavy metal data, the ANOVA two-factor with replication test ($\alpha = 0.01$) were applied to compare the algal and duckweed treatment ponds.

Knowing that in a continuous flow system there is always a variation in density and type of plants and microbes as the water passes through the system, the hydraulic model must account for these effects as reported by Kadlec and Wallace (2008). The tanks-in-series model (TIS) was applied and the removal profile for the contaminant in that system is given by:

$$\frac{\left(C-C^*\right)}{\left(C_i-C^*\right)}=\left(1+\frac{k\tau}{Nh}\right)^{-N} \qquad (1)$$

Where: C_i the initial concentration, C is the final concentration, C* is the background concentration, k is the first-order rate constant, N is the number of tanks in series, h is the depth and τ is the hydraulic retention time.

The mass balance analysis was performed for algae and duckweed systems as follows:

A) Duckweed system

$$Plant\ accumulation + Sedimentation = \sum\left(Q_{in}\times C_{in}\right)-\sum\left(Q_{out}\times C_{out}\right)-R \qquad (2)$$

Where: Q_{in}: influent flow rate (L/day); C_{in}: average influent concentration (mg/L); Q_{out}: effluent flow rate (L/day); C_{out}: average effluent concentration (mg/L); R: remaining load in water phase of the ponds at t_{final} (mg), hence R = total volume of the ponds × average conc. of last grab samples   (3)

B) Algal system

$$Algal\ accumulation + Biofilm\ accumulation = \sum\left(Q_{in}\times C_{in}\right)-\sum\left(Q_{out}\times C_{out}\right)-R \qquad (4)$$

4.3. Results

4.3.1. Tracer test

Tracer recovery was consistently 97 % or higher. Table 1 shows good agreement between theoretical and actual hydraulic retention times, indicating the absence of short-circuiting and / or dead zones. Dispersion numbers and N (number of tanks according to the tanks-in-series model) indicate well-mixed conditions.

Table 4-1: Hydraulic characteristics of the duckweed and the algal pond systems

Parameters	Eff_DP1	Eff_DP2	Eff_DP3	Eff_AP1	Eff_AP2	Eff_AP3
HRT_{theo} *(days)*	2.31	4.62	6.93	2.31	4.62	6.93
HRT_{act} *(days)*	2.29	4.81	6.90	2.27	4.88	6.92
Σ^2 *(hr^2)*	4006	17573	13112	4039	15624	13160
R_n	0.75	0.76	2.09	0.73	0.88	2.10
Dispersion number d	0.66	0.65	0.24	0.68	0.57	0.24
Dead zone *(%)*	0.87	-4.17	0.43	0.87	-5.52	0.43
Recovery *(%)*	100.00	100.00	100.00	100.00	97.60	97.60
Dead time *(hours)*	21.25	44.68	69.50	21.25	44.68	69.50

4.3.2. Physico-chemical parameters

Water temperature ranged between 19 – 22 °C (data not shown), with only small daily variations. Looking at the longitudinal profile, no significant differences (P > 0.05) in temperature values were found for all the duckweed ponds when compared to the algal ponds. Nevertheless, different temperature profiles were recorded when comparing DP to AP. In the AP, maximum and minimum temperatures were recorded, respectively, at 1:30 AM and 9:30 PM during Run 1 and Run 2. In Run 3, the temperature profile showed a constant profile. In the DP, a constant temperature profile was recorded during all runs.

During all runs, higher dissolved oxygen concentrations and pH values (Figures 4-3A & 4-4A, Table 4-2) were recorded in the algae ponds when compared to the duckweed ponds. However, the redox potential was high in the DP when compared to the AP (Figure 4-5B, Table 4-2). Statistical analysis revealed significant differences (t-test, P < 0.0001) between AP and DP in dissolved oxygen, pH and redox potential.

Table 4-2: Mean values of ORP, DO and pH for the AP and DP systems during all runs

	ORP (mV)			DO (mg/L)			pH		
	Run_1	Run_2	Run_3	Run_1	Run_2	Run_3	Run_1	Run_2	Run_3
DP1	262.5±46	332.5± 51.6	319±7.1	0.7±0.8	1.6±0.5	2.1±0.7	7.1±0.1	6.8±0.1	6.9±0.1
DP2	208±9.9	228± 12.7	362±28.3	3.3±0.4	2.2±1.0	1.3±0.8	6.9±0.1	6.6±0.2	7.0±0.1
DP3	285±12.7	335± 14.1	361±8.5	3.9±0.9	2.7±1.1	1.9±0.4	7.0±0.1	6.8±0.0	6.9±0.1
API	137±19.8	276± 90.5	232±15.6	16±4.8	8.6±2.4	14.3±0.8	9.4±0.4	8.4±0.7	9.1±0.1
AP2	153±43.8	271± 42.4	196.5±21.9	16.8±3.7	11.6±5.1	14.2±4.9	9.5±0.4	8.9±0.5	9.5±0.4
AP3	165±22.6	251± 31.1	186.5±9.2	11.8±2.9	10±2.9	16.3±1.6	9.2±0.3	8.9±0.3	9.8±0.1

1, 2 and 3 after AP and DP: stands for the number of the pond in series.

4.3.3. Metal removal

Table 3 gives an overview of influent and effluent concentrations based on four grab samples taken once the system has adapted to the influent composition. In general, looking at the longitudinal profile all five heavy metals were characterized by a decreasing pattern in their removal efficiencies when comparing Run 1, Run 2 and Run 3. Overall Cr and Zn showed the highest removal rates, both in the duckweed and algae ponds. The system performance was very good, varying between 91 – 99 % removal efficiency, both under low and high metal loading rates as well as under 24 hours of light regime. No significant differences were found between AP and DP in metal removal (ANOVA, P = 0.66) and after application of different light regimes (ANOVA, P = 0.89). However, the type of organisms did have a highly significant effect on the metal removal efficiency (ANOVA, P < 0.001) with algal ponds having a better performance.

Table 4-3: Influent and effluent concentrations (average ± standard deviation; n = 4) for different runs and ponds system.

	influent			Run_1						Run_2						Run_3					
	Run1	Run2	Run3	DP1	DP2	DP3	AP1	AP2	AP3	DP1	DP2	DP3	AP1	AP2	AP3	DP1	DP2	DP3	AP1	AP2	AP3
Cd (mg/L)	0.06 ± 0.01	0.13 ± 0.01	0.12 ± 0.00	0.053 ± 0.00	0.048 ± 0.00	0.040 ± 0.00	0.055 ± 0.01	0.053 ± 0.00	0.048 ± 0.00	0.118 ± 0.00	0.108 ± 0.00	0.098 ± 0.00	0.118 ± 0.00	0.110 ± 0.00	0.108 ± 0.01	0.118 ± 0.01	0.113 ± 0.01	0.105 ± 0.01	0.118 ± 0.01	0.115 ± 0.01	0.110 ± 0.000
Cr (mg/L)	1.35 ± 0.08	2.68 ± 0.15	2.68 ± 0.02	0.38 ± 0.04	0.18 ± 0.02	0.07 ± 0.01	0.19 ± 0.02	0.15 ± 0.01	0.05 ± 0.01	0.78 ± 0.15	0.25 ± 0.08	0.17 ± 0.02	0.53 ± 0.11	0.18 ± 0.10	0.06 ± 0.01	0.42 ± 0.25	0.21 ± 0.10	0.15 ± 0.05	0.37 ± 0.05	0.24 ± 0.51	0.07 ± 0.00
Cu (mg/L)	0.11 ± 0.01	0.24 ± 0.01	0.25 ± 0.01	0.1 ± 0.00	0.10 ± 0.01	0.08 ± 0.01	0.1 ± 0.01	0.09 ± 0.01	0.08 ± 0.01	0.21 ± 0.03	0.20 ± 0.01	0.20 ± 0.01	0.21 ± 0.02	0.20 ± 0.02	0.19 ± 0.02	0.2 ± 0.00	0.22 ± 0.01	0.19 ± 0.01	0.22 ± 0.01	0.20 ± 0.01	0.18 ± 0.01
Pb (mg/L)	0.26 ± 0.01	0.61 ± 0.01	0.54 ± 0.01	0.23 ± 0.01	0.19 ± 0.01	0.17 ± 0.01	0.24 ± 0.01	0.22 ± 0.01	0.21 ± 0.01	0.52 ± 0.03	0.48 ± 0.03	0.41 ± 0.01	0.54 ± 0.03	0.48 ± 0.02	0.44 ± 0.02	0.49 ± 0.01	0.44 ± 0.04	0.38 ± 0.02	0.48 ± 0.05	0.43 ± 0.05	0.36 ± 0.04
Zn (mg/L)	1.70 ± 0.05	3.14 ± 0.03	3.03 ± 0.02	1.25 ± 0.04	0.87 ± 0.05	0.48 ± 0.05	1.29 ± 0.03	0.94 ± 0.04	0.54 ± 0.09	2.82 ± 0.09	2.48 ± 0.16	2.25 ± 0.11	2.53 ± 0.18	2.25 ± 0.20	1.95 ± 0.32	2.59 ± 0.03	2.06 ± 0.07	0.65 ± 0.01	2.2 ± 0.07	1.49 ± 0.03	0.59 ± 0.05

1, 2 and 3 after AP and DP: stands for the number of the pond in series

4.3.4. Mass balance

The accumulation of heavy metals has been calculated for both duckweed and algal treatment ponds by the difference between inflows and outflows in the system including the quantity accumulated by the duckweed or algae, the quantity removed by sedimentation and the quantity remaining in the water phase of the ponds (Tables 4-4 and 4-5). The mass balance has been calculated for the entire experimental period, i.e. from the start up phase until the end of Run 3.

Table 4-4: Mass balance calculations for the duckweed system
(Load in mg / total operational time)

Metals Load (mg / 11 weeks)	Total Influent Load	Total Effluent load	Harvested duckweed	*R	Sedimentation
Pb	447 (100 %)	288 (64%)	101 (23%)	48 (11%)	10 (2 %)
Cd	98 (100 %)	75 (76%)	7 (7 %)	12 (12%)	5 (5 %)
Cr	2124 (100 %)	100 (5%)	1130 (53%)	17 (1%)	878 (41%)
Zn	2851 (100 %)	1173 (41%)	1123 (39%)	189 (7 %)	366 (13 %)
Cu	213 (100 %)	166 (78%)	20 (9%)	23 (11%)	5 (2 %)

*R: remaining load of water phase in the ponds

Table 4-5: Mass balance calculations for the algal system
(Load in mg / total operational time)

Metals Load (mg / 11 weeks)	Total Influent load	Total Effluent load	Floating algal layer	Final resident biofilm	*R	Error
Pb	447 (100 %)	280 (63 %)	10 (2 %)	108 (24 %)	42 (9 %)	7 (1.6 %)
Cd	98 (100 %)	78 (79 %)	1 (1 %)	6 (6 %)	12 (12 %)	2 (2 %)
Cr	2124 (100 %)	49 (2 %)	134 (6 %)	1924 (91 %)	12 (1 %)	6 (0.3 %)
Zn	2851 (100 %)	1090 (38 %)	45 (2 %)	1558 (55 %)	152 (5 %)	7 (0.2 %)
Cu	213 (100 %)	161 (75 %)	2 (1 %)	40 (19 %)	21 (10 %)	-10 (-4.7 %)

*R: remaining load of water phase in the ponds

In Table 4-4, it should be noted that sedimentation is presenting the accumulation of heavy metals at the bottom of the pond. In Table 4-5, the final resident biofilm is representing the layer of algae that was covering the walls of the algal ponds. Error is a closing term representing the difference in concentration found after summation of all components considered.

4.4. Discussion

4.4.1. General performance

Heavy metal concentrations in the effluents supplied to the experimental set-up were high compared to the standards reported in the WHO (2006) guidelines for the safe use of wastewater, excreta, and grey water and EPA (2009) drinking water contaminant limits. The following orders were observed in metal removal under different conditions: Cr > Zn> Pb > Cd > Cu in Run 1 for the duckweed system, Cr > Zn > Cu > Cd > Pb in Run 1 for the algal system, Cr > Zn = Pb > Cd = Cu in Run 2 for duckweed system, Cr > Zn> Pb > Cu > Cd in Run 2 for algae system and Cr > Zn > Pb > Cu > Cd in Run 3 for both systems. The first-order model for 2 tanks-in-series was fitted to our data and the rate constant k was determined. In general, the rate constant was low and followed the order Cr > Zn > Pb > Cu > Cd (Table 4-6).

Table 4-6: Rate constant (k) for the first order equation for the algal (AP) and the duckweed (DP) ponds

	k (/day)					
	Run 1		Run 2		Run 3	
	AP	DP	AP	DP	AP	DP
Cd	0.0151	0.0080	0.0103	0.0066	0.0047	0.0012
Cr	0.2310	0.2849	0.2023	0.3860	0.2198	0.3523
Cu	0.0120	0.0120	0.0065	0.0084	0.0100	0.1200
Pb	0.0160	0.0077	0.0150	0.0120	0.0130	0.0130
Zn	0.0600	0.0530	0.0085	0.0180	0.0790	0.0860

The system performance was good when compared with other reported natural systems. Maine et al. (2001) measured Cd uptake by Eichhornia crassipes, Hydromistia stolonifera, Pistia stratiotes and Salvinia herzogii. Little difference among species was observed and the approximate first order rate constant was 0.5 d^{-1}. Miretsky et al. (2004) reported on Cr removal by Salvinia herzogii and Pistia stratiotes. They found that the Cr uptake increases with increasing Cr concentration in water up to 6,200 mg/g at a water concentration of 6 mg/L. Murray-Gulde et al. (2005) reported an uptake of 6.7 % of the total Cu entering the system by Scirpus californicus. Removal of Pb in wetlands is highly variable with removal efficiencies between -220% and 98% for eight systems with a median removal of 25% (Kadlec and Wallace, 2008). Mitchell et al. (2002) reported a rate constant of 6/d for Zn in wetlands treating highway runoff.

Mass balance analyses showed that the duckweed has a high binding capacity when compared to algae, specifically for Cr, Zn and Pb. However, high metal concentrations were encountered in the final resident biofilm in the AP. This gives an indication of the mechanism of metal retention within the system. As we did not investigate the metal partitioning in the plant, it is difficult to precise which part of the plant played a major role in metal retention or accumulation. However, biosorption (Dhabab, 2011) and bio-uptake (Oporto et al., 2006; Hou et al., 2007) are reported heavy metal removal mechanisms. In the AP, it is clear that removal was occurring mainly via precipitation due to the high pH combined with biosorption and uptake. Our results are in good agreement with the findings of other researchers (Vymazal, 1995; Murray-Gulde et al., 2005; Vymazal and Krasa, 2005; Kadlec and Wallace, 2008; Sekomo et al., 2011a) who reported that algae, bulrushes, cattails, *Cyperus papyrus*, *Schoenoplectus californicus* play a lesser role in metal retention in a wetland. However, depending on the engineering process in the wetland, the bulk pollutant removal is always achieved by physico-chemical processes and then after plants can be used as polishing step.

4.4.2. Effect of metal load

The wastewater composition, in particular the dissolved oxygen concentration, the pH and the redox potential affect the speciation of heavy metals (Eggleton and Thomas, 2004; Simpson et al., 2004). In this study, physico-chemical parameters were affected as the metal load was varied (Table 4-2). Variation of pH between duckweed and algal ponds was related to the decreasing plant activities affected by doubling the heavy metal concentrations in Run 2. In Run 3, the increase in heavy metal removal observed was due to the application of a 24 hours light regime to the system. An increasing trend in redox potential was observed for both pond types in Run 2. That increase is directly linked to the decreasing DO influenced by higher metal loads in the system. In Run 3, a further increase was observed for the duckweed ponds.

Many wetland plants have constitutive metal tolerance and mechanisms of metal resistance (Matthews et al., 2005; Deng et al., 2006, 2009; Kanoun-Boule et al., 2009). These mechanisms enable most plant cells to continue normal activities upon Pb exposure while sacrificing a few cells that accumulate large amounts of Pb (Zhou et al., 2010). In this study, Figures 4 A and B, show the decrease in pH between run 1 and 2; corresponding to the doubling of the metal concentration. This pH decrease can be explained by a plant reaction to the metal concentration becoming toxic. This may result in the excretion of organic materials by plants and therefore decreasing the pH of the solution. Furthermore, Hou et al. (2007) reported on the way heavy metals enter frond chloroplasts and may be over-accumulated, thus inducing oxidative stress which causes damages like peroxidation of chloroplast membranes.

In Run 1, Zn exhibits the second best removal efficiency of all metals tested, reaching similar levels of around 70 % in both AP and DP (ANOVA, P = 0.36). Each of the three ponds in series seems to contribute to an almost equal way to the Zn removal, resulting in a linearly increasing removal efficiency. This linear trend could be explained by the same amount of pollutant being taken up by approximately equal amounts of duckweed in each pond. Important improvements can thus still be expected by enlarging the system; extrapolating the data shows that removal levels of 90 % and more could be reached by just adding one more pond.

For Run 2 the situation is quite different, with much lower overall Zn removal efficiencies: 28 % for DP and 38 % for AP. High metal concentrations thus seem to have a negative effect on Zn removal. Statistical analysis revealed significant differences between AP and DP (ANOVA, P < 0.001). Zn like Cu is known to be an essential micronutrient for normal plant metabolism, playing an important role in a large number of metalloenzymes, photosynthesis related plastocyanin and membrane structure (Megatelli et al., 2009, William et al. 2000). However, Cu and Zn are also known to be toxic heavy metals (Li and Xiong, 2004; Drost et al., 2007). Megatelli et al. (2009) reported Zn toxicity on *Lemna* at concentrations of 6.5 mg/L. The removal efficiency is still linearly increasing (about 10 % per pond). Extrapolating these data shows that one would have to triple the number of ponds to reach removal levels of 90 % and more.

In Run 3, a good removal efficiency for Zn was achieved: 82 % in AP and 79 % in DP. The light regime, stimulating the photosynthetic activity of the plants in the system, seems to be beneficial to the system (ANOVA, $P < 0.001$; Kruskal-Wallis, $P = 0.235$). An exponential increase in the removal efficiency was observed from pond 1 to pond 3 (Figure 4-2).

Cd, Cu and Pb all show relatively similar patterns with removal efficiencies varying between min. 17 % (Run 2: Cd in AP and Cu in DP) and max. 36 % (Run 1: Pb in DP) which indicates that both systems are not very suitable as a polishing step for removal of those heavy metals. ANOVA showed a statistically significant effect of the metal loading rate on Cu removal ($P < 0.05$) and of the pond type on Cd and Pb removal ($P < 0.01$ and $P < 0.001$, respectively). However, the Kruskal-Wallis test showed no statistically significant difference on the metal loading rate o Cu, Cd and Pb. The reason for such low removal could be explained by a toxicity effect affecting the plant performance. Megatelli et al. (2009) reported that Cd at 0.2 mg/L was the most toxic metal for *Lemna* followed by Cu at 0.6 mg/L and finally Zn at concentrations above 6.5 mg/L. Miretsky et al. (2004) also reported the non-survival of *Lemna* on a mixture of 4 mg/L of Zn, Cu, Fe, Mn, Cr and Pb each. Cd at concentrations above 1.12 mg/L and Cu at concentration above 3.18 mg/L promote pigment degradation and photosynthesis inhibition in *Lemna trisulca L* (Prasad et al., 2001). In this study, the low removal efficiency recorded in both systems could be also attributed to that toxic effect as the concentration used was close to the reported one.

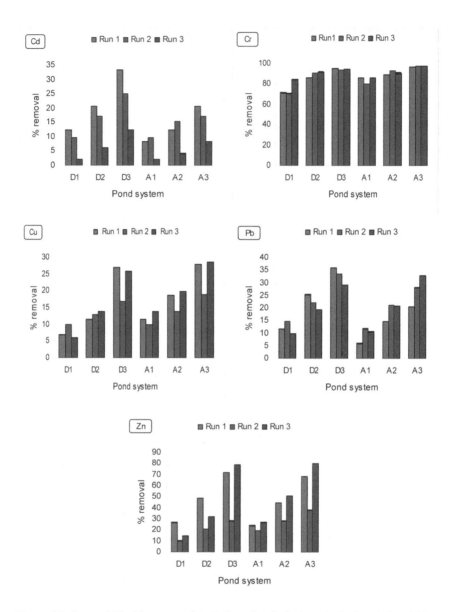

Figure 4-2: Removal % of heavy metal in duckweed and algal ponds for Run 1 (low HM concentration, 16 / 8 hours light regime), Run 2 (high concentration, 16 / 8 hours light regime) and Run 3 (high concentration 24 hours light regime), respectively.

Heavy metal toxicity to algae has been investigated for quite a long time. Many redox active and non redox reactive metals are known to cause oxidative stress, as indicated by lipid peroxidation and H_2O_2 accumulation in the cells (Schutzendubel et al., 2001). Copper is a redox-active transition metal, known to catalyze hydroxyl radical production. Furthermore, zinc and lead are redox inactive metals which inactivate the cellular antioxidant pool and disrupt the metabolic balance (Stohs and Bagchi, 1995; Briat, 2002). Shehata et al. (1999) reported on the toxic effect of multi metal mixtures which exist simultaneously in aquatic ecosystems on natural phytoplankton assemblages (green algae, blue-green algae and diatoms); They found substantial changes in phytoplankton community structure and the most tolerant group were the blue-green algae, followed by green algae while diatoms was the most sensitive group. These studies investigated toxicity effects of metal concentrations ranging between 0.05 and 0.2 mg/L of heavy metals. In our study, we investigated 5 to 10 times higher concentrations.

4.4.3. Effect of light

Exposure of the system to a 24 hours light regime has induced a change in physico-chemical parameters like DO, pH and redox potential. The sigmoid pattern recorded in Run 2 changed to a more or less constant pattern in Run 3 (Figures 4-3, 4-4 and 4-5). The reason for high oxygen levels in AP when compared to DP was due to algae photosynthetic activity within the water phase (Vymazal, 1995). In the DP, oxygen production occurred at the top surface of the plants (which is in the air phase) and oxygen tends to be lost to the atmosphere. Only a small portion is transported via the roots to the water phase (Bonomo et al., 1997). The duckweed mat at the top of the pond further prevents atmospheric re-aeration and light penetration hence limiting algae growth (Caicedo, 2005). In general, except Cd which showed high toxicity, the removal efficiencies of other metals have increased in run 3 when compared to run 2. This shows a positive effect of 24 hours light application to the system.

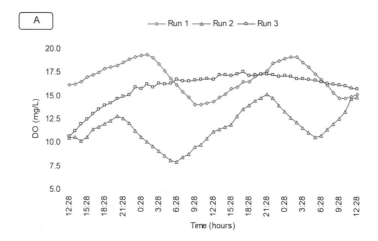

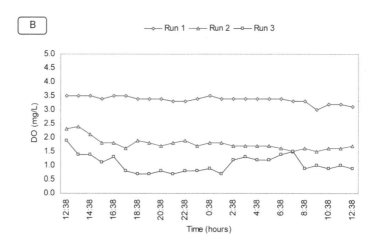

Figure 4-3: Typical DO profiles for algae (A) and duckweed (B) ponds for day / night regime during Run 1, Run 2 (doubling the heavy metal concentration) and Run 3 (24 hours light regime), respectively. Heavy metal concentrations 0.06, 1.35, 0.11, 0.26 and 1.7 mg/L in Run 1 and 0.12, 2.68, 0.25, 0.54 and 3.03 mg/L in Run 2 and 3 respectively for Cd, Cr, Cu, Pb and Zn.

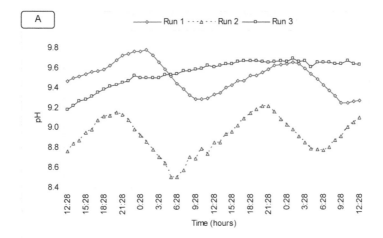

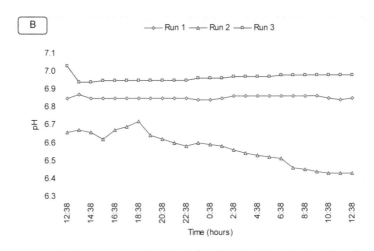

Figure 4-4: Comparison between pH profiles for algae (A) and duckweed (B) ponds for day / night regime during Run 1, Run 2 (doubling heavy metal concentration) and Run 3 (24 hours light regime), respectively. Influent pH was 6.98. Heavy metal concentrations 0.06, 1.35, 0.11, 0.26 and 1.7 mg/L in Run 1 and 0.12, 2.68, 0.25, 0.54 and 3.03 mg/L in Run 2 and 3 respectively for Cd, Cr, Cu, Pb and Zn.

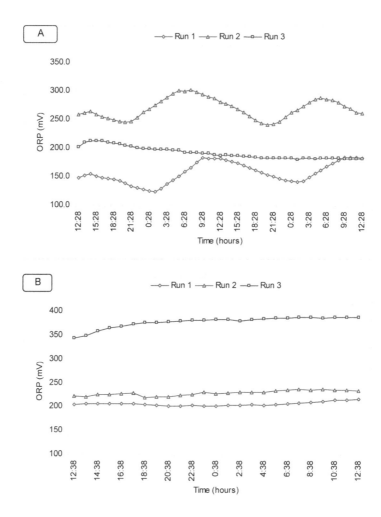

Figure 4-5: Redox profiles for algae (A) and duckweed (B) ponds for day / night regime during Run 1, Run 2 (doubling heavy metal concentration) and Run 3 (24 hours light regime), respectively. Heavy metal concentrations 0.06, 1.35, 0.11, 0.26 and 1.7 mg/L in Run 1 and 0.12, 2.68, 0.25, 0.54 and 3.03 mg/L in Run 2 and 3 respectively for Cd, Cr, Cu, Pb and Zn.

4.5. Conclusions

This study showed that duckweed and algal ponds are suited as polishing step for heavy metal removal at lower concentration. However, for the treatment of high pollutant loads a pre-treatment step is require to abatement of pollutant loads. Heavy metal removal and abiotic conditions (pH, DO, ORP) and light regimes did not yield differences in removal efficiency. The difference expected in metal removal based on the pond type was not confirmed. In general, the overall performance was close for both ponds.

Acknowledgments

The authors are grateful to the Dutch Government and the National University of Rwanda for the financial support provided for this research through the NPT/RW/051 Project. The authors would like also to thank the laboratory staff for UNESCO-IHE for their technical assistance during this research.

4.6. References

American Public Health Organisation, (1995). Standard Methods for the examination of water and wastewater, Water Environment Federation. Washington DC, USA.

Aziz, H.A., Adlan, M.N. & Ariffin, K.S. (2008). Heavy Metals (Cd, Pb, Zn, Ni, Cu and Cr(III)) removal from water in Malaysia: Post treatment by high quality limestone, Bioresource Technology 99(6), 1578 – 1583.

Blaylock, M.J. & Huang, J.W. (2000). Phytoextraction of metals. In: I. Raskin and B.D. Ensley eds. Phytoremediation of toxic metals: using plants to clean-up the environment. New York, John Wiley & Sons.

Bonomo, L., Pastorelli, G. & Zambon, N. (1997). Advantages and limitations of duckweed-based wastewater treatment systems. Water Science and Technology 35(5), 239 – 246.

Briat, J.F. (2002). Metal ion activated oxidative stress and its control. In: Inze, D., Montagu, M.V. (Eds.), Oxidative Stress in Plants. Taylor and Francis, New York, 171 – 189.

Brix, H. (1994). Functions of macrophytes in constructed wetlands. Water Science and Technology 29(4), 71 – 78.

Caicedo, B.J.R. (2005). Effect of operational variables on nitrogen transformations in duckweed stabilization ponds, PhD thesis, IHE Delft, The Netherlands.

Cohen, R.R.H. (2006). Use of microbes for cost reduction of metal removal from metals and mining industry waste streams, Journal of Cleaner Production 14(12-13), 1146 – 1157.

Cunningham, S.D., Shann, J.R., Crowley, D.E. & Anderson, T.A. (1997). Phytoremediation of contaminated water and soil. In: Kruger, E.L., Anderson, T.A. and Coats, J.R. eds. Phytoremediation of soil and water contaminants. ACS symposium series 664. Washington, DC, American Chemical Society.

Dhabab, J.M. (2011). Removal of Fe(II), Cu(II), Zn(II), and Pb(II) ions from aqueous solutions by duckweed. Journal of Oceanography and Marine Science. 2(1), 17-22

Deng, H., Ye, Z. & Wong, M. (2006). Lead and zinc accumulation and tolerance in populations of six wetland plants. Environmental Pollution 141, 69-80.

Deng, H., Ye, Z. & Wong, M. (2009). Lead, zinc and iron (Fe(II)) tolerances in wetland plants and relation to root anatomy and spatial pattern of ROL. Environmental and Experimental Botany 65, 353-362.

Drost, W., Matzke, M. & Backhaus, T. (2007). Heavy metal toxicity to Lemna minor: studies on the time dependence of growth inhibition and the recovery after exposure. Chemosphere 67, 36 – 43.

Eggleton, J. & Thomas, K.V. (2004). A review of factors affecting the release and bioavailability of contaminants during sediment disturbance events. Environment International 30, 973 – 980.

Ensley, B.D. (2000). Rational for use of phytoremediation. In: Raskin, I. and Ensley, B.D. eds. Phytoremediation of toxic metals: using plants to clean- up the environment. New York, John Wiley & Sons.

EPA, (2009). Online series. Drinking Water contaminants. Specific Fact Sheets for Consumer.http://www.epa.gov/safewater/contaminants/index.html. (Accessed 5 November 2009).

Galun, E., Siegel, B.Z., Keller, P., Lehr, H. & Siegel, S.M. (1987). Removal of metal ions from aqueous solutions by *Pencillium* biomass: Kinetic and uptake parameters, Water, Air and Soil Pollution 33, 359 – 371.

García, J., Chiva, J., Aguirre, P., Alvarez, E., Sierra, J.P. & Mujeriego, R. (2004). Hydraulic behaviour of horizontal subsurface flow constructed wetlands with different aspect ratio and granular medium size. Ecological Engineering 23, 177 – 187.

Hou, W., Chen, X., Song, G., Wang, Q. & Chi Chang, C. (2007). Effects of copper and cadmium on heavy metal polluted waterbody restoration by duckweed (Lemna minor). Plant Physiology and Biochemistry 45, 62 – 69.

Jacob, D.L. & Otte, M.L. (2004). Influence of Typha latifolia and fertilization on metal mobility in two different Pb–Zn mine tailings types. Science for the Total Environment. 333, 9 – 24

Johnson, D.B. & Hallberg, K.B. (2005). Acid mine drainage remediation options: a review. Science of the Total Environment 338, 3 – 14.

Kadlec, R.H. & Wallace, S.D. (2008). Treatment Wetlands second edition, CRC Press.

Kanoun-Boulé, M., Vicente, J.A.F., Nabais, C., Prasad, M.N.V. & Freitas, H. (2009). Ecophysiological tolerance of duckweeds exposed to copper. Aquatic Toxicology 91, 1 – 9.

Kelderman, P., Drossaert, W.M.E., Zhang, M., Galione, L., Okwonko, C. & Clarisse, I.A. (2000). Pollution assessment of the canal sediments in the city of Delft (The Netherlands). Water Research. 34, 936 – 944.

Kurniawan, T.A., Chan, G.Y.S., Lo, W.H. & Babel, S. (2006). Physico-chemical treatment techniques for wastewater laden with heavy metals. Chemical Engineering Journal 118, 83 – 98.

Lakatos, G., Kiss, M.K., Kiss, M. & Juhasz, P. (1997). Application of constructed wetlands for wastewater treatment in Hungary. Water Science and Technology 35(5), 331 – 336.

LeDuc, D.L. & Terry, N. (2005). Phytoremediation of toxic trace elements in soil and water. Journal of Industrial Microbiology and Biotechnology 32, 514 – 520.

Leman, M. (2000). Zinc removal by aquatic plants: Pistia stratiotes, MSc. Thesis, IHE Delft, The Netherlands.

Levenspiel, O. (1972). Chemical Reaction Engineering. 2nd ed., John Wiley & Sons, Inc., New York, NY.

Li, T.Y. & Xiong, Z.T. (2004). A novel response of wild type duckweed (*Lemna paucicostata Hegelm.*) to heavy metals. Environmental Toxicology 19, 95 – 102.

Maine M.A., Duarte M.V. & Sune N. (2001). Cadmium uptake by floating macrophytes. Water Research 35(11), 2629 – 2634.

Marchand, L., Mench, M., Jacob, D.L. & Otte, M.L. (2010). Metal and metalloid removal in constructed wetlands, with emphasison the importance of plants and standardized measurements: A review. Environmental Pollution 158, 3447 – 3461

Matthews, D.J., Moran, B.M. & Otte, M.L. (2005). Screening the wetland plant species Alisma plantago-aquatica, Carex rostrata, and Phalaris arundinacea for innate tolerance to zinc and comparison with Eriophorum angustifolium and Festuca rubra Merlin. Environmental Pollution 134, 343 – 351.

Miretsky, P., Saralegui, A. & Cirelli ,A.F. (2004). Aquatic Macrophytes potential for the simultaneous removal of heavy metals (Buenos Aires, Argentina). Chemosphere 57(8), 997 – 1005.

Murray-Gulde, C., Huddleston, G.M., Garber, K.V. & Rodgers, J.H. (2005). Contributions of Schoenoplectus californicus in a constructed wetland system receiving copper contaminated water. Water Air and Soil Pollution 163(1-4), 355 – 378.

Nriagu, J.O. (1979). Global inventory of natural and anthropogenic emissions of trace metals to the atmosphere. Nature 279, 409 – 411.

Oporto, C., Arce, O., Van den Broeck, E., Van der Bruggen, B. & Vandecasteele, C. (2006). Experimental study and modeling of Cr (VI) removal from wastewater using Lemna minor. Water Research 40, 1458 – 1464.

Prasad, M., Malec, P., Walaszek, A., Bojko, M. & Strzafka, K. (2001). Physiological responses of Lemna trisulca L. (duckweed) to cadmium and copper bioaccumulation. Plant Science 161, 881 – 889.

Qian, J.H., Zayed, A., Zhu, Y.L., Yu, M. & Terry, N. (1999). Phytoaccumulation of trace elements by wetland plants. Part III. Uptake and accumulation of ten trace elements by twelve plant species. Journal of Environmental Quality 28, 1448 – 1455.

Raskin, I., Kumar, P.B.A.N., Dushenkov, S. & Salt, D.E. (1994). Bioconcentration of heavy metals by plants. Current Opinion in Biotechnology 5(3), 285 – 290.

Rodgers, J.H.Jr. & Dunn, A. (1992). Developing design guidelines for constructed wetlands to remove pesticides from agricultural runoff. Ecological Engineering 1, 83 – 95.

Schutzendubel, A., Schwanz, P., Teichmann, T., Gross, K., Langenfeld-Heyser, R., Godbold, D.L. & Polle, A. (2001). Cadmium induced changes in antioxidative systems, hydrogen peroxide content, and differentiation in Scots pine roots. Plant Physiology 127, 887 – 898.

Shehata, S.A., Lasheen, M.R., Kobbia, I.A. & Ali, G.H. (1999). Toxic effect of certain metals mixture on some physiological characteristics of freshwater algae. Water Air and Soil Pollution. 110, 119 – 135.

Sheoran, A.S. & Sheoran, V. (2006). Heavy metal removal mechanism of acid mine drainage in wetlands: A critical review. Minerals Engineering 19, 105 – 116.

Simpson, S.L., Angel, B.M. & Jolley, D.F. (2004). Metal equilibration in laboratory-contaminated (spiked) sediments used for the development of whole-sediment toxicity tests. Chemosphere 54, 597 – 609.

Stohs, S.J. & Bagchi, D. (1995). Oxidative mechanisms in the toxicity of metal ions. Free Radical Bio Med 18, 321 – 336.

Schwarzenbach, R.P., Escher, B.I., Fenner, K., Hofstetter, T.B., Johnson, C.A., Von Gunten, U. & Wehrli, B. (2006). The Challenge of Micropollutants in Aquatic Systems. Science, 313, 1072 – 1075

Sekomo, C.B., Nkuranga, E., Rousseau, D.P.L. & Lens, P.N.L. (2011a). Fate of Heavy Metals in an Urban Natural Wetland: The Nyabugogo Swamp (Rwanda). Water Air and Soil Pollution 214, 321 – 333.

Sekomo, C.B., Kagisha, V., Rousseau, D.P.L. & Lens, P.N.L. (2011b). Heavy Metal Removal by Combining Anaerobic Upflow Packed Bed Reactors with Water Hyacinth Ponds. Environmental technology. DOI:10.1080/09593330.2011.633564.

Sharma, K.P., Sharma, S., Singh, P.K., Kumar, S., Grover, R. & Sharma, P.K. (2007). A comparative study on characterization of textile wastewaters (untreated and treated) toxicity by chemical and biological tests. Chemosphere 69(1), 48 – 54.

Shi, W., Wang, L., Rousseau, D.P.L. & Lens, P.N.L. (2010). Removal of estrone, 17α-ethinylestradiol, and 17β-estradiol in algae and duckweed-based wastewater treatment systems. Environmental Science and Pollution Research 17(4), 824 – 833.

Shilton, A. (2005). Pond Treatment Technology. London, UK, IWA Publishing. ISBN: 1843390205.

Megatelli, S., Semsari, S. & Couderchet, M. (2009). Toxicity and removal of heavy metals (cadmium, copper and zinc) by *Lemna gibba*. Ecotoxicology and Environmental Safety 72, 1774 – 1780.

Stigliani, W.M. (1992). Overview of the chemical time bomb problem in Europe. In: ter Meulen, G.R.B., Stigliani, W.M., Salomons, W., Brigdges, E.M., and Imeson, A.C. (eds) Proceedings of the European State-of-the-Art Conference on Delayed Effects of Chemicals in Soils and Sediments (Veldhoven, 2 – 5 September 1992). Foundation for Ecodevelopment. Hoofddorp, pp. 279.

Tang, S. (1993). Experimental study of a constructed wetland for treatment of acidic wastewater from an iron mine in China. Ecological Engineering 2, 253 – 259.

Vymazal, J. (1995). Algae and nutrient cycling in wetlands. CRC Press/Lewis Publishers: Boca Raton, Florida.

Vymazal, J. & Krasa P. (2005). Heavy metals budget for constructed wetland treatment municipal sewage. In: Natural and Constructed Wetlands – Nutrients, Metals and Management, Vymazal J. (ed) Backhuys Publishers: Leiden, The Netherlands, 135 – 142.

WHO, (2006). Guidelines for the safe use of wastewater, excreta and grey water. Volume 2 wastewater use in agriculture. http://www.who.int. (Accessed 5 November 2009).

Williams, L.E, Pittman, J.K. & Hall, J.L. (2000). Emerging mechanisms for heavy metal transport in plants. Biochimica et Biophysica Acta 1465 (12), 104 – 26.

Zimmo, O.R., Al Saed R. & Gijzen, H. (2000). Comparison between algae based and duckweed based wastewater treatment: differences in environment conditions and nitrogen transformations. Water Science and Technology 42 (10–11), 215 – 222.

Zimmo, O.R. (2003). Nitrogen transformations and removal mechanisms in Algal and Duckweed Wastewater Stabililization Ponds. PhD Dissertation, UNESCO – IHE Institute for Water Education and Wageningen University, The Netherlands

Zhou, Y.Q., Huang, S.Z., Yu, S.L., Gu, J.G., Zhao, J.Z., Han, Y.L. & Fu, J.J. (2010). The physiological response and sub-cellular localization of lead and cadmium in Iris pseudacorus L. Ecotoxicology 19, 69-76.

Chapter 5: Use of Gisenyi volcanic rock for adsorptive removal of Cd(II), Cu(II), Pb(II) and Zn(II) from wastewater

This chapter has been presented and published as:

Sekomo, C.B., Rousseau, D.P.L., & Lens, P.N.L., (2010). Use of Gisenyi volcanic rock for adsorptive removal of Cd(II), Cu(II), Pb(II) and Zn(II) from wastewater. In: *Proceedings of the 2nd International Conference Research Frontiers in Chalcogen Cycle Science and Technology.* UNESCO-IHE / Delft, The Netherlands (25 – 26 May 2010).

Sekomo, C.B., Rousseau, D.P.L., & Lens, P.N.L., (2012a) Use of Gisenyi Volcanic Rock for Adsorptive Removal of Cd(II), Cu(II), Pb(II), and Zn(II) from Wastewater. Water Air and Soil Pollution 223, 533 – 547.

Abstract

Volcanic rock is a potential adsorbent for metallic ions from wastewater. This study determined the capacity of Gisenyi volcanic rock found in Northern Rwanda to adsorb Cd, Cu, Pb and Zn using laboratory scale batch experiments under a variety of experimental conditions (initial metal concentration varied from 1 – 50 mg/L, adsorbent dosage 4 g/L, solid / liquid ratio of 1:250, contact time 120 hours, particle size 250 – 900 μm). The BET specific surface area of the adsorbent was 3 m^2/g. The adsorption process was optimal at near-neutral pH 6. The maximal adsorption capacity was 6.23 mg/g, 10.87 mg/g, 9.52 mg/g and 4.46 mg/g for Cd, Cu, Pb and Zn, respectively. The adsorption process proceeded via a fast initial metal uptake during the first 6 hours, followed by slow uptake and equilibrium after 24 hours. Data fitted well the pseudo second order kinetic model. Equilibrium experiments showed that the adsorbent has a high affinity for Cu and Pb followed by Cd and Zn. Furthermore, the rock is a stable sorbent that can be reused in multiple sorption–desorption–regeneration cycles, Therefore, the Gisenyi volcanic rock was found to be a promising adsorbent for heavy metal removal from industrial wastewater contaminated with heavy metals.

Key words: Volcanic rock, Rwanda, adsorption capacity, heavy metal, adsorption isotherms.

5.1. Introduction

Pollution of water bodies by heavy metals is a problem of environmental concern because it is threatening not only the aquatic ecosystem but potentially also human health through contamination of drinking water and food products (Cheng et al., 2002). Even though the toxic effect of heavy metals is known since long, water pollution by heavy metals has become acute in the past decades because of the rapid world industrialization (Tsezos, 2001). Many processes exist for removing dissolved heavy metals, including ion exchange, precipitation, phytoextraction, ultrafiltration, reverse osmosis, and electrodialysis (Applegate, 1984; Sengupta and Clifford, 1986; Geselbarcht, 1996; Schnoor, 1997). Conventional biological and chemical treatment methods are not cost effective for heavy metal removal, because of the non degradability properties of heavy metals. These processes may be very expensive, especially when the metals in solution are in the concentration range of 1-100 mg/L (Nourbakhsh et al., 1994). Another disadvantage of conventional treatment technologies is the production of toxic chemical sludge of which the disposal / treatment is rather costly.

Therefore, wastewater contaminated by heavy metals needs an effective and affordable technological solution, especially to treat the more dilute wastewater types such as acid mine drainage and industrial wastewaters that have metals in a concentration range of 1–10 mg/L, as those from the textile industry. Among the available techniques, adsorption has been used as one of the most practical methods. Adsorption is a process that utilizes low-cost adsorbents to sequester toxic heavy metals (Van Vliet et al., 1981; Slejko, 1985; Yang, 1999; Sharma, 2002). The major advantages of adsorption over conventional treatment methods include the utilization of low cost adsorbents and their high efficiency due to the possibility to regenerate the sorbent material. Therefore, it will avoid the production of solid toxic material as is the case for chemical sludge produced by conventional treatment technologies. It is important to stress here that, if the adsorbent is chosen carefully and the operational conditions are adjusted appropriately, adsorption processes are able to remove heavy metals over a wide pH range and to reach very low levels of heavy metals in the effluent compared to conventional techniques (Lai et al., 2002).

The nature of the adsorbent material is important because it plays a key role in the retention and accumulation of heavy metals from polluted water. Numerous types of adsorbent have been used for heavy metal accumulation, e.g. activated carbon, silica gel and activated alumina are popular and effective adsorbents, but their use is restricted because of their high costs (Davis, 1982; Trivedi and Axe, 2001; Mohan and Singh, 2002). Phosphate minerals or mineral apatites, respectively, hydroxyapatite and fluorapatite, natural zeolites and different types of clays have also been used successfully for the removal of heavy metals such as cadmium, chromium, lead and zinc in aqueous solutions by adsorption (Daza et al., 1991; Erdem et al., 2004; Ahmet et al., 2007; Murari et al., 2008). Iron oxide coated sand; a waste product from drinking water treatment plants shows a high adsorption capacity for Fe and Mn removal (Sharma et al., 1999; Buamah et al., 2008). Even if it is clear from the literature that considerable efforts have been made in finding suitable heavy metal adsorbents, there is still a need to orientate more efforts towards obtaining low-cost and efficient adsorbents that may be used in developing countries. Volcanic rocks may satisfy these requirements.

The most abundant volcanic rocks include pumice, a finely porous rock frothy with air bubbles and scoria, a rough rock that looks like furnace slag (Khandaker and Hossaim, 2004). These rocks are abundant in Southern Europe (Greece, Italy, Turkey and Spain), Central America, Southeast Asia and East Africa (Democratic Republic of Congo, Djibouti, Eritrea, Ethiopia, Kenya, Uganda and Rwanda) (Suh et al. 2008; Alemayehu and Lennartz, 2009). Recently, volcanic rocks have received considerable interest for heavy metal removal mainly due to their valuable properties: availability in many regions, high specific surface area and low cost. Their potential to remove heavy metals from polluted water bodies has been reported by among others Kwon et al. (2005) who showed that scoria (Jeju, Korea) has a larger capacity and affinity for Zn(II) sorption than commercial powdered activated carbon (PAC). Panuccio et al. (2008) studied cadmium-nutrient interactions with natural vermiculite, zeolite and pumice. Their results showed that the cadmium concentration in the nutrient (Hoagland) solution decreased with 30% when contacted with zeolite or vermiculite, but only to a minor extent when contacted with pumice. The amount of cadmium adsorbed on the mineral surfaces showed the following sequence zeolite > vermiculite > pumice.

The main objective of the present study was to characterize the Gisenyi volcanic rock from Northern Rwanda as adsorbent material, and to conduct adsorption studies with four selected heavy metals: Cd(II), Cu(II), Pb(II) and Zn(II). Experimental conditions were varied to identify the effect of pH, adsorbent dosage and increase in heavy metal concentration on the removal kinetics. The competition between the four selected heavy metals on their respective removal was also addressed. Finally, desorption studies were undertaken to check the regeneration capacity of this new adsorbent.

5.2. Materials and methods

5.2.1. Adsorbent characterization

In this study, volcanic rock was collected in Gisenyi (northern region of Rwanda), where all hills are only exclusively composed of these volcanic rocks. The rocks are brown with many pores and are very light compared to gravel or other rock types. The rock was crushed into small particles and sieved yielding a particle size between 250 and 900 µm used during all experiments.

5.2.1.1. Cation exchange capacity (CEC)

The CEC was determined via the ammonium acetate pH 7 leaching method (Essington, 2004). A rock sample was extracted with an initial volume of ammonium acetate, followed by a leaching phase with ethanol to remove occluded NH_4^+. Once the leachate is free of NH_4^+, the soil is leached with KCl. The CEC in kg equals the concentration of displaced NH_4^+ in the leachate divided by the mass of soil in the column.

5.2.1.2. Bulk density

A measuring cylinder was filled with pulverized volcanic rock to a certain volume, and then the mass of this volcanic rock was weighed on a balance. The average bulk density ρ_b [g/L] was calculated (Blake and Hartge, 1986) according to:

$$\rho_b = \frac{m_{v.r}}{V_1} \qquad (1)$$

Where:
$m_{v.r}$: mass of volcanic rock occupying a volume V_1 in a measuring cylinder

5.2.1.3. Particle density

A measuring cylinder was filled with water to a certain volume, and then a weighed mass of volcanic rock was added to the water in the cylinder. The mix was left standing for 24 hours to allow the trapped air to escape. The total volume was measured the next day and the average particle density ρ_p [g/L] was calculated (Ruhlmanna et al. 2006) according to:

$$\rho_p = \frac{m_{v.r}}{V_2} \qquad (2)$$

Where:
$m_{v.r}$: mass of volcanic rock added to the water in the cylinder and V_2: volume of volcanic rock calculated as volume of water subtracted from total volume.

5.2.1.4. Porosity

Porosity η [%] was calculated from average bulk and particle densities (Blake and Hartge, 1986):

$$\eta = \left(1 - \frac{\rho_b}{\rho_p}\right) \times 100 \qquad (3)$$

5.2.1.5. Point of zero charge

The point of zero charge (PZC) was determined by acid-base potentiometric titration. A weighted mass of the pulverized rock was dispersed by stirring in a cell containing $NaNO_3$ supporting electrolyte and the whole solution was purged by bubbling purified nitrogen gas. Successive titrations of the solution have been conducted by addition of NaOH 0.1 M. The pH range investigated was 3 to 11. The titration was performed in duplicate at three different ionic strengths 10^{-1}, 10^{-2} and 10^{-3} mol of $NaNO_3$. The common intersection point of the three titration curves was identified as the PZC (De Faria et al., 1998; Appel et al., 2003).

5.2.1.6. Chemical composition and elemental analysis.

The chemical composition of the adsorbent was determined by X–Ray Fluorescence (XRF) spectrometry. The samples were pressed into powder tablets and the measurements were performed with Philips PW 2400 WD – XRF spectrometer, the data evaluation was done with Uniquant software. The elemental analysis was conducted on 0.5 g of pulverized rock (particle size: < 250 μm) was mixed with 10 mL of concentrated nitric acid into a digestion tube. Thereafter, the tube was closed and placed in a CEM Mars 5 Microwave Accelerated Reaction System. The digestion was conducted at high pressure (10 bar). After the digestion, the sample was diluted and the elemental analysis was conducted on an ICP-AES Perkin Elmer Optima 3000.

5.2.1.7. Surface area

The surface area has been determined according to the BET (Brunaeur, Emmet and Teller) theory. This method is applicable to non-porous, meso-porous and macroporous materials (Gregg and Sing, 1992). The sample cell holding the outgassed sample is evacuated and cooled to liquid nitrogen temperature (77 K). Portions of nitrogen are dosed into the sample cell and will be partly adsorbed on the surface, eventually getting into equilibrium with the gas phase. In that way adsorption and desorption points can be recorded at different pressures, typically in the relative pressure range between 0.05 and 0.25, resulting in an ad- and desorption isotherm.

5.2.2. Adsorption tests

5.2.2.1. Experimental set-up

Adsorption tests were done by putting 1 g of adsorbent in contact with 250 mL of a heavy metal solution (pH 6.0 ± 0.1) at a concentration of ± 1 mg/L, unless specified otherwise. Concentrations of Cd, Cu, Pb and Zn ions in the initial solutions were prepared by dissolving the appropriate weight of the respective salts $Cd(NO_3)_2 \times 4H_2O$, $Cu(NO_3)_2 \times 2H_2O$, $Pb(NO_3)_2$, $Zn(NO_3)_2 \times 6H_2O$ in doubly distilled water. The anions of these salts (NO_3^-) in the aqueous solutions do not form any metal ion complexes, and their effect on the ion exchange process is considered to be nihil. The flasks were capped to avoid pH changes during experiments due to CO_2 escape. Also, during the experimentation period, few drops of 0.1M HNO$_3$ or NaOH were added to the flasks for pH readjustment where required. An average variation of 0.3 pH unit was observed after 5 days of experimentation. Each day a sample was taken after 1 minute settling time and filtered on 0.45 µm Whatman filters. Then pH was measured with a Metrohm 781 pH/ion meter and the samples were acidified with a drop of 65 % HNO$_3$ prior to heavy metal analysis on a flame atomic absorption Perkin Elmer model AAnalyst 200. The amount of heavy metals adsorbed at time t, q$_t$ and the absorbed percentage [A (%)] was calculated from the equation:

$$q_t = \frac{(C_o - C_t) \times V}{M} \tag{4}$$

$$A(\%) = \left(\frac{C_o - C_t}{C_o}\right) \times 100 \tag{5}$$

Where C_o = initial concentration of heavy metals in contact with adsorbents (mg/L), q_t = the amount of heavy metals adsorbed per unit mass of the adsorbent (mg/kg), M = mass of the adsorbent (kg), C_t = mass concentration of heavy metal in aqueous phase at time t (mg/L), V = initial volume of the aqueous phase in contact with the adsorbents during the adsorption test (L), A (%) = adsorbed amount given as percentage at time t_t (%).

5.2.2.2. Effect of pH on adsorption

The effect of pH on the adsorbent was assessed by putting 1g of adsorbent in contact with 250 mL of heavy metal solution at a concentration of 1 mg/L. A control solution was also used by putting the same volume of the metal solution in a flask without the adsorbent. The pH was varied from 4 to 7, if necessary few drops of 0.1 M HNO$_3$ or NaOH were added to the metal solution to fix the value when variation was observed.

5.2.2.3. Effect of adsorbent on pH

The effect of the adsorbent on the pH was assessed by putting 1g of adsorbent in contact with 250 mL of distilled water. pH changes were observed when the adsorbent was kept in contact with distilled water solution for 24 and 120 hours at a fixed initial pH value using few drops of 0.1 M HNO$_3$ or NaOH solutions.

5.2.2.4. Adsorption kinetics and isotherm determination

The data necessary for fitting the kinetic model and the sorption isotherms were obtained in batch experiments as described above. The method used was the batch equilibrium technique (Ho, 1995; Ho et al., 1995; Ho and Ofomaja, 2006). For the determination of adsorption kinetics and isotherms, 250 mL of heavy metals solutions of 1, 10, 25 and 50 mg/L were shaken with an adsorbent dose varying from 0.64 to 4 g/L on a shaker at 250 rpm for 120 h at room temperature (25°C). All experiments were done in duplicate for isotherm establishment and in triplicate for kinetic studies.

5.2.2.5. Effect of competing ions

Equilibrium experiments were conducted to investigate the influence of the competition between the four metals on their uptake onto the adsorbent. Initial equimolar concentrations (40 mmol/L) for each metal were used. The adsorption of each metal under competitive and noncompetitive conditions was investigated by contacting 250 mL of a four metal mixture or a single metal solution with 4 g/L adsorbent. The pH was adjusted to 6 ± 0.1 and solutions were then put on an orbital shaker at 250 rpm.

5.2.3. Desorption experiments

Desorption experiments were conducted to investigate the reusability and regeneration potential of the adsorbent. The same adsorbent was used in consecutive adsorption–desorption (cycles I, II and III) or adsorption–desorption–regeneration (cycles IV and V). In each cycle, the adsorbent was filtered and repeatedly treated with 250 mL of 0.1M HNO$_3$ for 6 h and washed with deionized water after each desorption cycle to eliminate the excess of acid. In the regeneration step, the adsorbent was soaked in 250 mL of 1M KCl solution for 12 h, filtered, and washed with deionized water before being reused in a new adsorption–desorption cycle.

5.2.4. Adsorption kinetics and sorption isotherms

5.2.4.1. Theoretical background on adsorption kinetics

The mechanism of adsorbate sorption onto an adsorbent can be described by several mathematical models. The parameters found in these expressions are important in water and wastewater treatment process design. The kinetics of heavy metals adsorption were analyzed using two different kinetic models: the pseudo first order and the pseudo second order. The pseudo first order equation describing the kinetic process of liquid-solid phase adsorption is given by the general expression (Lagergren, 1898):

$$\frac{dq_t}{dt} = k_1 \left(q_e - q_t \right) \qquad (6)$$

Where q_e and q_t are the adsorption capacities at equilibrium and time t, respectively (mg/kg), k_1 is the pseudo first order rate sorption constant (1/h). After integration and applying the boundary conditions $q_t = 0$ at $t = 0$, $q_t = q_e$ when $t = t_e$, the linear form of the equation (6) becomes:

$$\log \left(q_e - q_t \right) = \log q_e - k_1 t \qquad (7)$$

When the removal of a solute follows a two step process, consisting of (1) sharp increase of the solute removal during the initial contact time between the solution and solid phase followed (2) a slow increase until an equilibrium state is reached, it can be represented by a pseudo second order kinetic model (Ho and McKay, 1999; Ho, 2006):

$$\frac{dq_t}{dt} = k_2 \left(q_e - q_t \right)^2 \qquad (8)$$

Where k_2 is the pseudo second order rate sorption constant (kg/mg.h), which represents the steepness of the curve (as in the case of Figure 3). After rearrangement, integration and applying the boundary conditions, the linear form of the equation (8) becomes:

$$\frac{t}{q_t} = \frac{1}{V_o} + \frac{1}{q_e} t \qquad (9)$$

and $V_o = k_2 q_e^2$ (10)

Where V_o (mg/kg.h) means the initial adsorption rate, and the constant k_2 can be determined experimentally by plotting t/q_t against t.

5.2.4.2. Theoretical background on adsorption isotherms

In general, the adsorption process is assessed by establishment of the adsorption isotherms (Ho, 1995, Ho et al., 1995). The Langmuir and Freundlich models are the most frequently used ones to describe the adsorption isotherms from experimental data. These models provide information to predict the removal efficiency of solute and an estimation of adsorbent amounts needed to remove solute ions from aqueous solution. Their mathematical expressions are:

Freundlich: $\quad q_e = K C_e^{1/n}$ (11)

Langmuir: $\quad q_e = \dfrac{abC_e}{1+bC_e}$ (12)

Where q_e (mg / kg) is the specific amount of heavy metal adsorbed, and C_e (mg/L) is the heavy metal concentration in the liquid phase at equilibrium. The parameter K (L/kg) and 1/n are the Freundlich constants that are related to the total adsorption capacity and intensity of adsorption, respectively. The Langmuir parameters are a (mg/kg), which is related to adsorption density, and b (L/mg), which is indicative of the adsorption energy. From the Langmuir equation a further analysis can be made on the basis of a dimensionless equilibrium parameter, R_L (Babu and Gupta, 2008), also known as the separation factor. It is given by the expression:

$$R_L = \dfrac{1}{1+bC_e}$$ (13)

The value of R_L lies between 0 and 1 for favorable adsorption; $R_L > 1$ represents unfavorable adsorption; $R_L = 1$ represents linear adsorption; and the adsorption process is irreversible if $R_L = 0$. The constant b (L/mg) denotes the adsorption energy. When the initial section of an isotherm is nearly linear, the slope it defines is called the distribution coefficient K_d (L/kg). This coefficient is the ratio between the content of the substance in the solid phase and its concentration in the aqueous solution, when adsorption equilibrium is reached. This coefficient is used to characterize the mobility of heavy metals: low K_d values imply that most of the metal remains in solution, while high K_d values indicate that the metal has great affinity to the adsorbent (Morera et al., 2001). This coefficient can be determined using the following equation:

$$Q_e = K_d C_e$$ (14)

5.3. Results and discussion

5.3.1. Chemical and physical properties of volcanic rock

The XRF analysis revealed SiO_2 is the dominant component, comprising of 43.24 %. Other components contributing more than 10 % are Al_2O_3 (16.65 %), Fe_2O_3 (14.15 %) and CaO (10.95 %) (Table 5-1). Results from this study showed that the chemical composition of the Gisenyi volcanic rock is similar to the chemical composition of volcanic scoria with SiO_2 47.4 %, Al_2O_3 (21.6 %), Fe_2O_3 (8.9 %) and CaO (12.4 %) (Alemayehu and Lennartz, 2009).

Table 5-1: Chemical composition of porous volcanic rock determined by XRF

Components	%
SiO_2	43.24
Al_2O_3	16.65
Fe_2O_3	14.15
CaO	10.95
MgO	4.29
TiO_2	3.57
Na_2O	2.44
K_2O	2.35
P_2O_5	0.77
SO_3	0.43
BaO	0.40
MnO	0.23
Others	0.53

Tables 2 and 3 present the elemental composition and physical properties of the volcanic rock studied. The composition analysis showed that the mineral structure already contains Cu and Zn (Table 5-2). The presence of heavy metals in volcanic rock has also been reported by Petalas et al. (2006) who studied the hydrochemistry of waters of volcanic rocks. They found the following order of dominance for the heavy metals in surface and ground waters: Fe > Mn > Zn > Ni > Cu. The multi-point BET surface area results revealed that the sample displays a sufficiently high surface area of 3 m^2/g to achieve a maximum measurement accuracy in accordance with ISO 9277 (Demiral et al., 2008).

Table 5-2: Elemental analysis of volcanic rock (values in mg/g)

Al	Ba	Ca	Cd	Cr	Cu	Fe	K	Mg	Mn	Na	Ni	Pb	Zn
11.95	0.25	12.10	<0.05	<0.05	1.20	26.50	5.65	11.55	0.53	2.45	<0.05	<0.5	0.96
±	±	±			±	±	±	±	±	±			±
0.21	0.00	0.57			0.00	0.71	0.35	0.21	0.01	0.21			0.01

Table 5-3: Physical properties of pulverized volcanic rock

Bulk density, ρ_b [g/cm^3]	CEC [meq/100g]	Particle density, ρ_p [g/cm^3]	pzc [pH]	Porosity, η [%]	S_{BET} [m^2/g]
1.25 ± 0.04	2.40 ± 0.22	3.76 ± 0.02	7.20 ± 0.12	66.70 ± 0.24	3.00 ± 0.00

5.3.2. Effect of adsorbent on the pH

Table 4 shows the pH variation during 24 and 120 hours contact time with the adsorbent. The pH variation during the experiments was between 0 and 1.4 unit on average at low and near neutral pH (pH 4 – 6), this can be interpreted in terms of proton adsorption. However, at high pH (pH 8 – 10), an average variation between 0 and 1.0 was observed. At pH 11 and 12 there were only small changes. This behavior could be attributed to the reaction of hydroxide ions with the adsorbent surface or to protons release from surface groups reacting with the solution. Similar observations were also reported by Alemayehu and Lennartz (2009) in their investigation on volcanic scoria and pumice.

Table 5-4: Effect of adsorbent on the pH of solution after 24 and 120 hours exposure time.

pH initial	4	5	6	7	8	9	10	11	12
pH (24 h)	4.89	5.89	6.46	7.11	7.30	7.58	9.85	10.92	12.03
pH (120 h)	5.49	6.74	7.03	7.25	7.44	7.64	8.94	10.78	11.92

5.3.3. Effect of contact time on the adsorption process

The effect of the contact time on the adsorption of Cd(II), Cu(II), Pb(II) and Zn(II) on volcanic rock is shown in Figure 5-1. Initially sorption takes place very quickly and then it continued slowly up to the maximum sorption. The equilibrium was reached after 24 hours and thereafter the amount of adsorbed metal ions did not change significantly with an increase in contact time. Therefore, a contact time of maximum 24 hours could be considered as an optimum time for adsorption of Cd(II), Cu(II), Pb(II) and Zn(II).

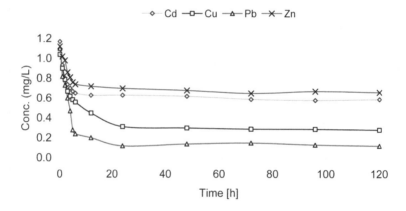

Figure 5-1: Effect of contact time on heavy metal adsorption onto volcanic rock. Cd(II): ◊, Cu(II): ▢, Pb(II): Δ and Zn(II):X. Test conducted at pH 6 with initial heavy metal concentration of 1 (± 0.2) mg / L and 4 g / L of adsorbent mass.

As reported by Babu and Gupta (2008), the mechanism of solute transfer to the solid includes diffusion through the layer or fluid film around the adsorbent particle and diffusion through the pores to the internal adsorption sites. Initially the concentration gradient between the layer or film and the solid surface is large, and hence the transfer of solute onto the solid surface is faster. That is why it takes only a short time to attain a high removal percentage of these heavy metals. As time increases, intraparticle diffusion becomes dominant. Hence solute takes more time to transfer from the solid surface to internal adsorption sites through the pores. Our findings are in good agreement with what has been reported by other researchers (Alemayehu and Lennartz, 2009; Kacaoba, 2009 and Demiral et al., 2008), i.e. a quick initial uptake of heavy metals by the adsorbent and reaching the equilibrium between 6 and 24 hours of contact time.

5.3.4. Effect of initial heavy metal concentration

The amount of heavy metal being adsorbed per unit mass of adsorbent increased with an increasing heavy metal concentration of the solution. When the initial concentration of the solution was changed from 1 to 50 mg/L, the quantity of heavy metal adsorbed increased from 230 to 6000; 270 to 10290; 260 to 6460; and 210 to 5401 mg/kg of adsorbent respectively for Cd, Cu, Pb and Zn. This observation has also been reported by Alemayehu and Lennartz (2009) in their research using volcanic scoria and volcanic pumice for Cd removal. They noticed an increase in the adsorbed quantity as the metal solution was more concentrated.

5.3.5. Effect of pH on the adsorption process

The pH of the aqueous solution is a significant parameter for removal of metal ions by adsorption (Essington, 2004; Jenne 1998, Faust and Aly, 1998; Martell and Hancock, 1996) but is also affecting precipitation. In the present study, a modeling analysis using PHREEQC software was conducted to check at which pH values precipitation was likely to occur in the solution and, given the focus on adsorption, experiments were only conducted in a pH region where metal precipitation is not important. The results from PHREEQC (data not shown) showed positive saturation indices for Cu and Pb precipitates at pH 7. However, for Zn and Cd they were positive at pH 8 and 9, respectively. For that reason, our adsorption experiments were conducted at pH ≤ 7 to prevent the possible precipitation of $Cd(OH)_2$, $Cu(OH)_2$, $Pb(OH)_2$ and $Zn(OH)_2$, which would introduce uncertainty into the interpretation of the results.

Figure 2 shows that adsorption onto volcanic rock is highly pH dependant. In acidic solution around pH 4 no adsorption of Cd(II), Cu(II) and Zn(II) was observed; Pb(II) however did exhibit adsorption at pH 4. That behavior of Pb can be explained by the competition between hydrated metal ions $[M^{+n}(H_2O)_n]$ and proton ions $[H^+]$ on the interface of the adsorbent. One may conclude that the hydrated Pb ions have been preferentially adsorbed onto the adsorbent surface rather than the proton ions. This observation is supported by Faust and Aly (1998), who reported that hydrous oxides of Fe(III) adsorb Pb(II) over the pH range of 4.3 to 7.0. Table 2 and 3 show the chemical composition and elemental analysis where iron is among the major species contained in the rock.

At pH 6, all four metals were adsorbed on the adsorbent. It is clear that near-neutral pH is favorable to the adsorption process. This is mainly due to displacement of the water equilibrium towards the predominance of $[HO^-]$ ions. The decrease of $[H^+]$ in aqueous solution as competing ions with metal ions on the adsorption sites is favoring the adsorption of metal ions by electrostatic attraction, whereas $[HO^-]$ has been repelled from the adsorption sites. The increasing adsorption potential of the media with increasing pH is a trend which is in agreement with the findings of Buamah et al. (2008) using iron oxide coated sand for As and Mn removal; Sari et al. (2007) using Celtek clay for Cr(III) and Pb removal; Pehlivan et al. (2009) using dolomite powder to remove Pb and Cu.

In order to verify whether or not the decrease of the heavy metal concentration could be due to the precipitation of the metal ions in the forms of metal hydroxides, a control experiment was each time conducted on the metal solution without adsorbent. In Figure 2, when comparing the control and the test curves from pH 4 to 7, it is clear that the lowering of heavy metal concentration was mainly due to the adsorptive properties of the rock rather than by precipitation and glass wall adsorption which accounted for less than 12.5 % for Cd, 9.1 % for Cu, 15.9 % for Pb and 2.8 % for Zn at pH 6 (Figure 5-2). Furthermore, the modeling analysis with PHREEQC showed that precipitation accounted for 3.7 % for Cu, 0.066 % for Pb and 0 % for Cd and Zn respectively at pH 6. The decrease pattern in metal concentration observed in the control could be explained by precipitation combined with adsorption onto the wall of the container as the flasks used were made of glass. Essington (2004), reported on a species distribution diagram of hydrolytic metal species in a solution as a function of pH where it appears that for all studied heavy metals precipitation is only expected to occur above pH 7.

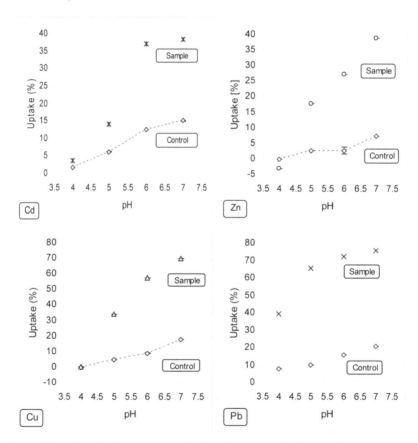

Figure 5-2: Effect of pH on heavy metal adsorption. The dashed and solids lines represent variation of the uptake in the control and the samples respectively. Test conducted with 1 (± 0.2) mg / L of each heavy metal concentration (n = 3), contact time: 120 h and 4 g / L adsorbent mass.

5.3.6. Adsorption kinetics

The adsorption kinetics of the four selected heavy metals have been studied for a contact time ranging from 1 to 120 hours by monitoring the percentage removal of heavy metals by the adsorbent. Table 5-5 reports the equilibrium parameters q_e (mg/kg) and the adsorption rate constant k_2 (kg/mg.h), the calculated equilibrium capacity values ($q_{e,calc}$) were close to the experimentally determined values (q_e). The equilibrium capacity value ($q_{e, calc}$) was extracted from the regression line in Figure 5-3 representing the pseudo second-order sorption kinetics model. In general, application of the pseudo first order kinetic model resulted in very bad fits for all heavy metals with an R^2 factor ranging between 0.31 and 0.42. However, excellent fits ($R^2 > 0.99$) were obtained for all heavy metals when the pseudo-second order kinetic model was applied. This model implies that the rate limiting step may be chemical adsorption involving valent forces through sharing or exchange of electrons between the adsorbent and the heavy metal ions (Ho and McKay, 1999; Ho and McKay, 2000).

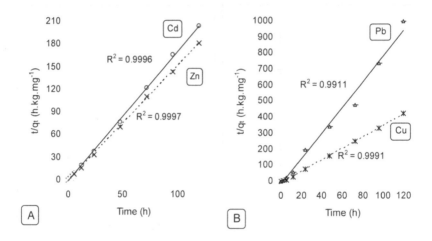

Figure 5-3: Plots of pseudo second-order sorption kinetics model. Heavy metal concentration 1 mg / L; contact time: 120 h; 4 g / L adsorbent mass and pH 6 (± 0.1).

Table 5-5: Values of pseudo second-order rate kinetic constants

heavy metal	q_e (mg/kg)	$q_{e, calc.}$ (mg/kg)	k_2 (kg/mg.h)	R^2
Cd	580	585	2.278	0.9996
Cu	281	279	0.002	0.9991
Pb	120	125	3.103	0.9911
Zn	650	658	2.226	0.9997

5.3.7. Adsorption isotherms

Figure 4 shows the isotherms plots of the four metals. Data from Table 5-6 did not show any significant differences (Anova, $P > 0.05$) between the R^2 from Langmuir and Freundlich models in the concentration range investigated. Similarly, other adsorption studies conducted on various adsorbents have also reported that both Freundlich and Langmuir isotherms fit well their data (Alemayehu and Lennartz, 2009; Kacaoba, 2009; Sari et al., 2007). This is somewhat surprising because the two adsorption models have different basic assumptions: the Langmuir model is based on the formation of a monolayer whereas the Freundlich model is based on the formation of multilayer on the adsorbent (Jenne, 1998). The fact that the Freundlich equation agrees well with the Langmuir equation is most probably because of the moderate concentration ranges used in this study (1 – 50 mg/L), as also shown by McKay et al. (1980) and Allen et al. (2004).

Furthermore, the Freundlich and Langmuir theories suffer from the disadvantage that equilibrium data over a wide concentration range cannot be fitted with a single set of constants. This could potentially be explained by the fact that the adsorption process starts in the range where many adsorption sites are only beginning to be filled by the metal ions, the relationship between K_d (distribution coefficient) and $C_{M,aq}$ (concentration of the metal in the solution) may be linear with a near zero slope. Then, after a small fraction of the sites are filled with metal ion adsorbed (M_j), the relationship may be linear but with a decrease in slope. K_d falls off more quickly due to a reduced number of available adsorption sites unfilled with M (metal ion), increasing competition from the desorbed ions, and a decreasing adsorbate concentration where $C_{Mj, aq}$ is constant (Jenne, 1998).

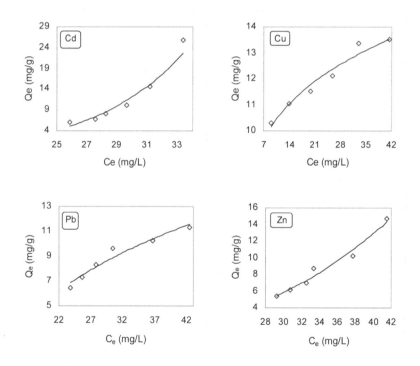

Figure 5-4: Adsorption isotherms of Cd, Cu, Pb and Zn. Test conducted at pH 6 (± 0.1), metal concentration was 50 (± 0.2) mg/L, adsorbent dose 4 g/L, agitation 250 rpm.

The maximum adsorption capacity X_m followed the order Cu > Pb > Cd > Zn. The constant b indicating the adsorption or desorption energy was different for each metal ion, indicating differences in the retention intensity. The high values of correlation coefficient indicate a good agreement between the parameters governing the adsorption of Cd(II), Cu(II), Pb(II) and Zn(II). Furthermore, the dimensionless constant R_L remained between $0 < R_L < 1$, consistent with the requirement for a favorable adsorption process.

Table 5-6: Freundlich and Langmuir isotherm constants of the Gisenyi volcanic rock.

Metal Solution	Freundlich constants			Langmuir constants				
	K (L/mg)	1/n	R^2	X_m (mg/g)	b (L/mg)	R^2	R_L	K_d
Cd	2.037	0.946	0.974	6.289	0.381	0.986	0.962	1.876
Cu	5.828	0.942	0.992	10.869	0.657	0.993	0.976	12.988
Pb	13.554	1.051	0.981	9.524	1.265	0.973	0.974	28.500
Zn	1.647	0.918	0.985	4.464	0.463	0.981	0.952	1.695
10 mg/L								
Cd	10^{-4}	0.197	0.983	0.477	0.117	0.976	0.591	0.183
Cu	6.450	0.594	0.958	9.174	0.393	0.989	0.829	4.696
Pb	1.427	0.702	0.962	10.526	0.123	0.976	0.859	1.583
Zn	10^{-3}	0.311	0.938	0.380	0.077	0.962	0.077	0.077
25 mg/L								
Cd	0.074	2.974	0.931	6.094	0.115	0.926	0.688	1.358
Cu	2.654	0.444	0.943	12.136	0.116	0.913	0.670	1.228
Pb	0.244	1.837	0.956	12.674	0.056	0.953	0.782	1.019
Zn	2×10^{-4}	3.735	0.891	2.336	0.044	0.853	0.643	0.250
50 mg/L								
Cd	3×10^{-8}	5.839	0.954	2.368	0.027	0.985	0.589	0.232
Cu	14.108	0.156	0.886	14.306	0.265	0.899	0.296	1.149
Pb	0.036	1.643	0.915	13.850	0.014	0.935	0.749	0.270
Zn	5×10^{-4}	2.769	0.969	4.849	0.018	0.959	0.655	0.185

The Freundlich adsorption capacity K followed the trend Pb > Cu > Cd > Zn. The adsorption intensity $1/n$ less than one is an indication of favorable adsorption, as new sites could be available and the adsorption capacity would be increased (Table 5-6). For values exceeding one (Table 5-6), this was an indication of weak adsorption bonds. This shows that the capacity of the adsorbent is decreasing because of its saturation as reported by Aziz et al. (2008) and Alumayehu and Lennartz (2009). Furthermore, the distribution coefficients (K_d) were calculated for both sets of concentrations. The result showed that Pb and Cu have higher K_d values (28.50 and 12.99 L/kg, resp.) than Cd and Zn (1.88 and 1.69 L/kg, resp.) for low metal concentrations. For high metal concentrations, the K_d values were 1.15 and 0.27 for Cu and Pb, while it was 0.23 and 0.18 for Cd and Zn, respectively (Table 5-6). This reflects the high affinity that Pb and Cu have for volcanic rock and its adsorption capacity at low and high metal concentrations.

Table 5-7: Reported adsorption capacities (mg/g) for several sorbents

adsorbents	pH	Particle size (μm)	Adsorbent dosage (g/L)	Metal conc. (mg/L)	Adsorption capacity (mg/g)				Reference
					Cd	Cu	Pb	Zn	
Ball clay	4.0 – 6.5	250	20	20 – 100	2.24	1.60	–	2.88	Ulmanu et al. (2003)
Bentonite	6.0	430 – 1000	20	65 – 200	9.27	18.16	–	–	Pehlivan et al. (2009)
Diatomite	6.0	430 – 1000	20	65 – 200	3.24	5.54	–	–	Pehlivan et al. (2009)
Dolomite	2.0 – 7.0	75 – 177	10	2.5 – 66	–	8.26	21.74	–	Hong et al. (1999)
Kaolinite	6.0	430 – 1000	10	1 – 100	3.04	4.47	–	–	Pehlivan et al. (2009)
Phosphate rocks	2.0 – 4.0	150 – 212	5	10 – 500	–	9.70	9.40	5.00	Prasad et al. (2008)
Sepiolite	0.0 – 3.0	–	2.5 – 25	100	0.39	–	–	–	Kocaoba (2009)
Scolecite	5.0 – 6.0	62 – 250	5	5 – 30	0.18	4.20	5.80	2.10	Bosso and Enzweiler (2002)
Volcanic pumice	6.0	75 – 425	4 – 50	2 – 50	3.84	–	–	–	Alemayehu et al. (2009)
Volcanic Scoria	6.0	75 – 425	4 – 50	2 – 50	2.24	–	–	–	Alemayehu et al. (2009)
Natural Zeolite	5.5 – 6.5	100 – 500	10	30 – 1,850	–	25.04	–	13.06	Peric et al. (2004)
Gisenyi volcanic rock	6.0	250 – 900	0.64 – 4	1 – 50	6.29	14.31	13.85	4.85	This study

Table 5-7 summarizes reported adsorption capacities to evaluate sorbent effectiveness. Sorption depends heavily on experimental conditions such as pH, metal concentration, sorbent concentration, competing ions, and particle size. Therefore, the reported sorption capacities are indications of values that can be achieved under a specific set of conditions rather than as maximum sorption capacities. Results of this study showed very good maximum adsorption capacity of Cd, Cu, Pb and Zn onto the adsorbent in general when compared to the other sorbents (Table 5-7).

5.3.8. Removal efficiency

The equilibrium concentration of the heavy metal investigated decreases with the increase of the adsorbent dosage. Figure 5-5 shows that all heavy metals have been characterized by increasing removal efficiency at higher adsorbent dosage. Cd(II) was characterized by a removal efficiency from 60 – 90 %; Cu(II) from 80 – 97 %; Pb(II) from 89 – 98 % and finally Zn(II) from 53 – 88 % by increasing the adsorbent dosage from 0.64 to 4 g / L. At the same time, a decreasing trend in adsorption capacity was observed while the adsorbent quantity increased. The adsorption capacity dropped from 1 – 0.24 mg / g for Cd(II); 1.38 – 0.27 mg / g for Cu(II); 1.44 – 0.25 mg / g for Pb(II) and 0.78 – 0.21 mg / g for Zn(II). This is mainly due to the sites remaining unsaturated during the adsorption process. The increase was not linear when the adsorbent dosage increases. This could be explained by the heterogeneity of the adsorbent with a particle size between 250 – 900 µm.

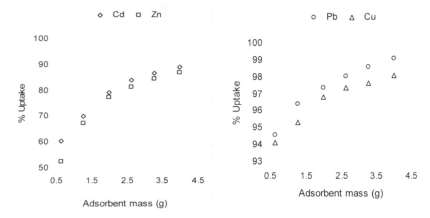

Figure 5-5: Removal efficiency of heavy metals as a function of mass of adsorbent. Test conducted at pH 6 (± 0.1) with metal concentration of 1 (± 0.2) mg/L.

5.3.9. Effect of competing ions

The metal removal efficiency under competitive conditions was approximately three times lower for Cd and Zn, i.e. from 85.8 to 29.3 % and from 84.7 to 30.8 % respectively. For Cu and Pb, there was less than 10% reduction in the removal efficiency, i.e. from 96.0 to 86.2 % and from 97.3 to 89.5 %, respectively. This indicates that metal ions are competing for the same adsorbing sites of the volcanic rock. Depending on the adsorption conditions, the effect of competing ions will also vary as the adsorption of metal ions is known to be governed by the free metal concentration, the nature and the quantity of the adsorbents (Alemayehu and Lennartz, 2009; Chantawong et al., 2003).

5.3.10. Desorption

The desorption experiment recovers the metal retained and the adsorbent can be reused in subsequent loading cycles. Different types of eluent agents have been reported (Atdor et al., 1995; Zhou et al., 1998; Ahuja et al., 1999), mainly inorganic acids, organic solutions and complexing agents (Hashim et al. 2000; Gong et al., 2005). In this work, Figure 5-6 shows the adsorption capacity and the performance of desorption and regeneration processes of the volcanic rock.

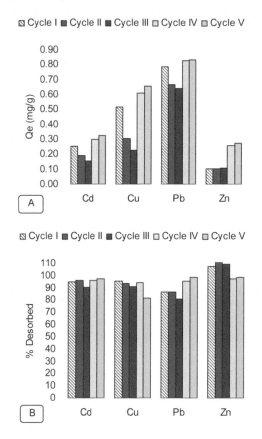

Figure 5-6: Metal uptake (A) and desorption performance (B) of volcanic rock after each adsorption–desorption cycle with 0.1M HNO₃ (cycles I, II & III) and adsorbent regeneration step with 1M KCl (cycles IV & V).

Figures 5-6A and 5-6B show the consecutive adsorption and desorption cycles. Without regeneration of the rock, the adsorption capacity of the sorbent was decreasing from 0.25 to 0.16 mg/g for Cd, from 0.52 to 0.23 mg/g for Cu and from 0.79 to 0.64 for Pb, respectively (cycles I, II and III). However, for Zn it stayed more or less constant. With regeneration using a 1M KCl solution (cycles IV and V), the sorbent adsorption capacity increased from 0.16 to 0.33 mg/g for Cd, from 0.23 to 0.66 mg/g for Cu, from 0.64 to 0.84 for Pb and from 0.10 to 0.27 for Zn, respectively. The 8 to 15 % increase in Zn concentration observed can be explained by the dissolution of the existing heavy metal in the crystalline structure of the rock (Table 5-2).

The use of KCl in the regeneration step was beneficial to the recovery of Pb and Zn. However, for Cu a slight decrease was observed in the last cycle (Figure 5-6). As reported by other researchers, Hong et al., (1999); Beolchini et al., (2003); Sekhar et al., (2004); Hammaini et al. (2007); Mata et al., (2010); Katsou and Tzanoudaki, (2010) the regeneration can be conducted with EDTA or with other solutions like CaCl$_2$ or KCl reacting as ion exchanger to avoid the dissolution of the heavy metals from the mineral structure,. As strong complexing agent, EDTA recovers the adsorbed metal from the adsorbent surface by complexation reactions at one side and at the other side, ions will replace the metal at their bonding site by cations exchange, both mechanisms are not as aggressive as strong acids to the rock structure.

5.4. Conclusions

This study has demonstrated the capacity of Gisenyi volcanic rock as an effective adsorbent for heavy metal removal from aqueous solution. The removal efficiency is affected by parameters investigated such as contact time, pH, adsorbent dosage, heavy metal concentration and competition with other heavy metals. The adsorption increases with pH and adsorbent dosage. The dimensionless adsorption constant R_L agrees well with the condition supporting favorable adsorption. The pseudo second-order kinetic model fitted very well the experimental data and therefore suggests a removal mechanism of chemical nature. The adsorption proceeded initially by a fast metal uptake followed by a slow and more or less constant uptake. Parameters from the adsorption isotherms indicated that the adsorbent was able to adsorb Pb and Cu ions to a large extent when compared to Cd and Zn. Removal of Cd, Cu Pb and Zn from polluted water by adsorption onto volcanic rocks as alternative low cost and abundant adsorbent seems therefore technically feasible. Furthermore, the volcanic rock is a stable sorbent that can be reused in multiple sorption–desorption–regeneration cycles, using 0.1M HNO$_3$ and 1M KCl as desorbing and regenerating agents respectively.

Acknowledgements

The authors are grateful to the Dutch Government and the National University of Rwanda for the financial support provided for this research through the NPT / RW / 051- WREM Project.

5.5. References

Ahmet, S., Mustafa, T. & Mustafa, S. (2007). Adsorption of Pb(II) and Cr(III) from aqueous solution on Celtek clay. Journal of Hazardous Materials 144, 41 – 46.

Ahuja, P., Gupta, R. & Saxena, R.X. (1999). Sorption and desorption of cobalt by Oscilatoria anguistissima. Current Microbiology 39, 49 – 52.

Alemayehu, E. & Lennartz, B. (2009). Virgin volcanic rocks: Kinetics and equilibrium studies for the adsorption of cadmium from water. Journal of Hazardous Materials 169, 395 – 401.

Allen, S.J., McKay, G. & Porter, J.F. (2004). Adsorption isotherm models for basic dye adsorption by peat in single and binary component systems. Journal of Colloid and Interfarce Science 280, 322 – 333

Appel, C., Ma, L..Q., Rhue, R.D. & Kennelley, E. (2003). Point of zero charge determination in soils and minerals via traditional methods and detection of electroacoustic mobility. Geoderma 113, 77– 93.

Applegate, L.E. (1984). Membrane separation processes, Chemical Engineering. 91, 64 – 89.

Atdor, I., Fourest, E. & Volesky, B. (1995). Desorption of cadmium from algal biosorbent. Canadian Journal of Chemical Engineering 73, 516 – 522.

Aziz, H.A., Adlan, M.N. & Ariffin, K.S. (2008). Heavy metals (Cd, Pb, Zn, Ni, Cu and Cr (III)) removal from water in Malaysia: post-treatment by high quality limestone, Bioresource Technology, 99 (6), 1578 – 1583.

Babu, B.V. & Gupta, S. (2008). Adsorption of Cr(VI) using activated neem leaves: kinetic studies, Adsorption 14, 85 – 92.

Beolchini, F., Pagnanelli, F., Toro, L. & Veglio, F. (2003). Biosorption of copper by Sphaerotilus natans immobilised in polysulfone matrix: equilibrium and kinetic analysis, Hydrometallurgy 70, 101–112.

Blake, G.R. & Hartge, K.H. (1986). Bulk Density, in A. Klute, ed., Methods of Soil Analysis, Part I. Physical and Mineralogical Methods: 2nd ed. Madison, American Society of Agronomy, (Agronomy Monograph, 9), 363 – 375.

Bosso, S.T. & Enzweiler, J. (2002). Evaluation of heavy metal removal from aqueous solution onto scolecite. Water Research 36 (19), 4795 – 4800.

Buamah, R., Petruseveski, B. & Schippers, J.C. (2008). Adsorptive removal of manganese(II) from the aqueous phase using iron oxide coated sand. Journal of Water Supply. Research and Technology 57, 1 – 11.

Chantawong, V., Harvey, N.W. & Bashkin, V.N. (2003). Comparison of heavy metal adsorptions by Thai kaolin and ballclay. Water, Air, and Soil Pollution, 148, 111 – 125.

Cheng, J., Bergamann, B.A., Classen, J.J., Stomp, A.M. & Howard, J.W. (2002). Nutrient recovery from swine lagoon water by Spirodela punctata. Bioresource Technology 81, 81– 85.

Davis, J.A. (1982). Adsorption of natural dissolved organic matter at the oxide / water interface, Geochimica et Cosmochimica Acta 46, 2381 – 2393.

Daza, L., Mendioroz, S. & Mendioroz, J.A. (1991). Mercury adsorption by sulfurized fibrous silicates, Clays Clay Minerals 39, 14 – 21.

De Faria, L.A., Prestat, M., Koenig, J.F., Chartier, P. & Trasatti, S. (1998). Surface properties of Ni + Co mixed oxides: a study by X-rays, XPS, BET and PZC. Electrochimica Acta 44, 1481 – 1489.

Demiral, H., Demiral, I., Tumsek, F. & Karabacakoglu, B. (2008). Adsorption of chromium(VI) from aqueous solution by activated carbon derived from olive bagasse and applicability of different adsorption models. Chemical Engineering Journal 144 (2), 188 – 196.

Erdem, E., Karapinar, N. & Donat, R. (2004). The removal of heavy metal cations by natural zeolites. Journal of Colloid and Interface Science 280, 309 – 314.

Essington, M.E. (2004). Soil and Water Chemistry: An integrative Approach, CRC Press, Washington, USA.

Faust, S.D. & Aly, O.M. (1998). Chemistry of water treatment 2nd edition, CRC Press LLC, USA.

Geselbarcht, J. (1996). Micro Filtration / Reverse Osmosis Pilot Trials for Livermore, California, Advanced Water Reclamation, in: Water Reuse Conference Proceedings, WWA 187.

Gong, R., Ding, Y., Lui, H., Chem, Q. & Liu, Z. (2005). Lead biosorption and desorption by intact and pretreated spirula maxima biomass. Chemosphere, 58, 125 – 130.

Gregg, S.J. & Sing, K.S.W. (1992). Adsorption, surface Area and Porosity, 2nd ed., Academic Press, London, UK.

Hammaini, A., Gonzalez, F., Ballester, A., Blazquez, M.L. & Munoz, J.A. (2007). Biosorption of heavy metals by activated sludge and their desorption characteristics. Journal of Environmental Management 84, 419 – 426.

Hashim, M.A., Tan, H.N. & Chu, K.H. (2000). Immobilized marine algal biomass for multiple cycles of copper adsorption and desorption. Separation and Purification Technology 19, 39 – 42.

Ho, Y.S. & McKay, G. (1999). Pseudo-second order model for sorption processes, Process Biochemistry 34, 451 – 465.

Ho, Y.S. & McKay, G. (2000). The kinetics of sorption of divalent metal ions onto sphagnum moss peat. Water Research 34 (3), 735 – 742.

Ho, Y.S., Wase, D.A.J. & Forster C.F. (1995). Batch nickel removal from aqueous solution by sphagnum moss peat. Water Research 29, 1327 - 1332.

Ho, Y.S. & Ofomaja, A.E. (2006). Pseudo-second-order model for lead ion sorption from aqueous solutions onto palm kernel fiber. Journal of Hazardous Materials B129, 137–142

Ho, Y.S. (1995). Adsorption of Heavy Metals from Waste Streams by Natural Materials, PhD Thesis, The University of Birmingham, UK.

Ho, Y.S. (2006). Review of second-order models for adsorption systems. Journal of Hazardous Materials 136 (3), 681 – 689.

Hong, P.K.A., Li, Ch., Banerjiand, S.K. & Regmi, T. (1999). Extraction, recovery and biostability of EDTA for remediation of heavy metal contaminated soil. Journal of Soil Contamination 8, 81 – 103.

ISO 9277 – Determination of the specific surface area of solids by gas adsorption – BET method.

Jenne, E.A. (1998). Adsorption of metals by geomedia. Variables, mechanisms, and model applications. Academic Press, California, USA.

Katsou, E. & Tzanoudaki, M. (2010). Regeneration of natural Zeolite polluted by lead and zinc in wastewater treatment systems, Journal of Hazardous Materials, doi:10.1016/j.jhazmat.2010.12.061

Khandaker, M. & Hossaim, A. (2004). Properties of volcanic pumice based cement and lightweight concrete, Cement and Concrete Research 34, 283 – 291.

Kocaoba, S. (2009). Adsorption of Cd(II), Cr(III) and Mn(II) on natural sepiolite. Desalination 244, 24 – 30.

Kwon, J.S., Yun, S.T., Kim, S.O., Mayer, B. & Hutcheon, I. (2005). Sorption of Zn (II) in aqueous solution by scoria, Chemosphere 60 (10), 1416 – 1426.

Lagergren. S. (1898). About the theory of so-called adsorption of solution substances. Kungliga Svenska Vetenskapsakademiens. Handlingar, Band 24(4), 1 – 39.

Lai, C.H., Chen, C.Y., Wei, B.L. & Yeh, S.H. (2002). Cadmium adsorption on goethite coated sand in the presence of humic acid, Water Research 36 (20), 4943 – 4950.

Mata, Y.N., Blazquez, M.L., Ballester, A., Gonzalez, F. & Munoz. J.A. (2010). Studies on sorption, desorption, regeneration and reuse of sugar-beet pectin gels for heavy metal removal. Journal of Hazardous Materials 178, 243 – 248

Martell, A.E. & Hancock, R.D (1996). in: J.P. Fackler Jr. (Ed.), Metal Complexes in Aqueous Solutions, Plenum Press, New York, USA.

McKay, G., Otterburn, M.S. & Sweeney, A.G. (1980) Kinetics of colour removal from effluent using activated carbon. Journal of the Society of Dyers and Colourists 96, 576 – 579.

Mohan, D. & Singh, K.P. (2002). Single and multi-component adsorption of cadmium and zinc using activated carbon derived from bagasse — an agricultural waste. Water Research 36, 2304 – 2318.

Morera, M.T., Echeverria, J.C., Mazkiaran, C. & Garrido, J.J. (2001). Isotherms and sequential extraction procedures for evaluating sorption and distribution of heavy metals in soils. Environmental Pollution 113, 135 – 144.

Murari, P., Huan-yan, X. & Sona, S. (2008). Multi-component sorption of Pb(II), Cu(II) and Zn(II) onto low-cost mineral adsorbent, Journal of Hazardous Materials 154, 221 – 229.

Nourbakhsh, M., Sag, Y., Ozer, D., Aksu, Z., Katsal, T. & Calgar, A. (1994). Comparative study of various biosorbents for removal of chromium (VI) ions from industrial wastewater. Process Biochemistry 29, 1 – 5.

Panuccio, M.R., Crea, F., Sorgona, A. & Cacco, G. (2008). Adsorption of nutrients and cadmium by different minerals: experimental studies and modelling, Journal of Environmental Management 88 (4), 890 – 898.

Pehlivan, E., Müjdat, O.A., Dinc, S. & Parlayici, S.E. (2009). Adsorption of Cu^{2+} and Pb^{2+} ion on dolomite powder, Journal of Hazardous Materials 167, 1044 – 1049.

Peric, J., Trgo, M. & Medvidovic, NV. (2004). Removal of zinc, copper, and lead by natural Zeolite: a comparison of adsorption isotherms. Water Research 38 (7), 1893 – 1899.

Petalas, C., Lambrakis, N. & Zaggana, E. (2006). Hydrochemistry of waters of volcanic rocks: the case of the volcano sedimentary rocks of Thrace, Greece. Water, Air, and Soil Pollution 169, 375 – 394

Prasad, M., Xu, H.Y. & Saxena, S. (2008). Multi-component sorption of Pb(II), Cu(II) and Zn(II) onto low-cost mineral adsorbent. Journal of Hazardous Materials 154, 221 – 229.

Ruhlmanna, J., Korschensb, T.M. & Graefe, J. (2006). A new approach to calculate the particle density of soils considering properties of the soil organic matter and the mineral matrix. Geoderma 130, 272 – 283.

Sarı, A., Tuzen, M. & Soylak, M. (2007). Adsorption of Pb(II) and Cr(III) from aqueous solution on Celtek clay. Journal of Hazardous Materials 144, 41 – 46

Schnoor, J.L. (1997). Phytoremediation, Technology Evaluation Report TE-97-01, Ground-Water Remediation Technologies Analysis Center, Pittsburgh, PA, USA.

Sekhar, K.C., Kamala, C.T., Chary, N.S., Sastry, A.R.K., Rao, T.N. & Vairamani, M. (2004). Removal of lead from aqueous solutions using an immobilized biomaterial derived from a plant biomass, Journal of Hazardous Materials 108, 111–117.

Sengupta, A.K. & Clifford, D. (1986). Important process variables in chromate ion exchange, Environmental Science and Technology 20, 149 – 155.

Sharma, S.K. (2002). Adsorptive Iron Removal from Groundwater. PhD Thesis. UNESCO-IHE Institute for Water Education and Wageningen University. Swets and Zeitlinger B.V., Lisse, The Netherlands.

Sharma, S.K., Greetham, M.R. & Schippers, J.C. (1999). Adsorption of iron(II) onto filter media. Journal of Water Supply: Research and Technology - Aqua, 48 (3), 84 – 91.

Slejko, F.L. (1985). Adsorption Technology–A step-by-step Approach to Process Evaluation and Application. In: Chemical Industries Series Vol. 19. F.L. Slejko, (ed). Tall Oaks Publishing Inc. Voorhees, New Jersey, USA 18 – 30.

Suh, C.E., Luhr, J.F. & Njome, M.S. (2008). Olivine-hosted glass inclusions from Scoriae erupted in 1954 – 2000 at Mount Cameroon volcano, West Africa. Journal of Volcanology and Geothermal Research 169 (1–2), 1 – 33.

Trivedi, P. & Axe, L. (2001). Predicting divalent metal sorption to hydrous Al, Fe and Mn oxides, Environmental Science and Technology 35 (9), 1779 – 1784.

Tsezos, M. (2001). Biosorption of metals. The experience accumulated and the outlook for technology development. Hydrometallurgy 59, 241–243.

Ulmanu, M., Maranon, E., Fernandez, Y., Castrillon, L., Anger, I. & Dumitriu, D. (2003). Removal of copper and cadmium ions from diluted aqueous solutions by low cost and waste material adsorbents. Water, Air, and Soil Pollution 142, 357 – 373.

Van Vliet, B.M., & Weber Jr., W.J. (1981). Comparative performance of synthetic adsorbents and activated carbon for specific compound removal from wastewaters. Journal of the *Water Pollution Control* Federation. 53, *(*11), 1585 – 1598.

Yang, R.T. (1999). Gas Separation by Adsorption Processes.–Series on Chemical Engineering, Vol. 1. Publishers–Imperial College Press, London, UK.

Zhou, J.L., Huang, P.L. & Lin, R.G. (1998). Sorption and desorption of Cu and Cd by macroalgae and microalgae. Environmental Pollution 101, 67 – 75.

Appendix 5.1

Potentiometric determination of the point of zero charge

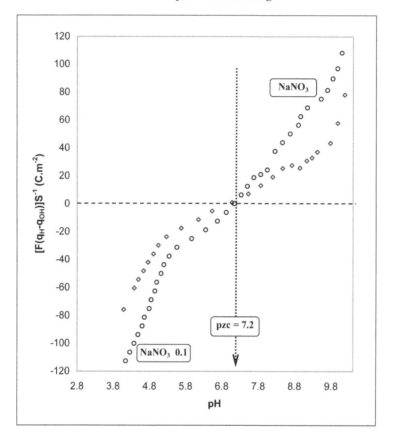

Appendix 5.2

Photo of the porous volcanic rock

Chapter 6: Heavy Metal Removal by Combining Anaerobic Upflow Packed Bed Reactors with Water Hyacinth Ponds

This chapter has been presented and published as:

Sekomo, C.B., Rousseau, D.P.L., & Lens, P.N.L., (2009). Heavy Metal removal by a combined system of anaerobic Upflow Packed Bed Reactor and Water Hyacinth Pond. In: *Proceedings of the Ecological Engineering; from Concepts to Application (EECA)*. Cité Internationale Universitaire de Paris, France (02 – 04 December 2009).

Sekomo, C.B., Kagisha, V., Rousseau, D.P.L., & Lens, P.N.L., (2011b) Heavy Metal Removal by Combining Anaerobic Upflow Packed Bed Reactors with Water Hyacinth Ponds. Environmental Technology. DOI:10.1080/09593330.2011.633564

Abstract

The removal of four selected heavy metals (Cu, Cd, Pb and Zn) has been assessed in an Upflow Anaerobic Packed Bed reactor filled with porous volcanic rock as adsorbent and attachment surface for bacterial growth. Two different feeding regimes were applied using low (5 mg/L of heavy metal each) and high (10 mg/L of heavy metal each) strength wastewater. After a start up and acclimatization period of 44 days, each regime was operated for a period of 10 days with a hydraulic retention time of 1 day. Good removal efficiencies of at least 86 % have been achieved for both low and high strength wastewater. The bioreactor performance was not much affected when the columns were operated under high strength heavy metal concentration. A subsequent water hyacinth pond with a hydraulic retention time of 1 day removed an additional 61 % Cd, 59 % Cu, 49 % Pb and 42 % Zn, showing its importance as polishing step. The water hyacinth plant in the post treatment step accumulated heavy metals mainly in the root system. Overall metal removal efficiencies at the outlet of the integrated system were 98 % for Cd, 99 % for Cu, 98 % for Pb and 84 % for Zn. Therefore, the integrated system can be used as an alternative treatment system for metal-polluted wastewater for developing countries.

Key words: Heavy metals, Industrial wastewater, Sulphate reducing bacteria, Upflow Anaerobic Packed Bed reactor, Volcanic rocks.

6.1. Introduction

Large quantities of water are used in the textile industry for cleaning the raw material as well as for many flushing steps during the whole production. The resulting wastewater has to be cleaned from fat, oil, colour and other chemicals, which are used during the various production steps. Until now, most attention has been given to the removal of dyes (Banat et al. 1996; Juang et al. 1996; Fu and Viraraghavan, 2001; Marcucci et al. 2002; Ozdemira et al. 2009). Few studies have reported on heavy metal removal of textile wastewater (Srivastava and Majumder, 2008; Ali et al. 2009).

Several physico-chemical methods have been used for decades in heavy metal removal from industrial wastewater, such as ion exchange, activated carbon, chemical precipitation, chemical reduction and filtration (Beszedits, 1988; Pansini et al. 1991; Perez et al. 1995, Rengaraj et al. 2001). These conventional methods present however some limitations: they are expensive and can themselves generate other waste problems due to the intensive use of chemicals. Considerable interest has also been developed to address metal removal via adsorption onto different kind of adsorbents (Daza et al., 1991; Erdem et al., 2004; Ahmet et al. 2007, Murari et al. 2008, Suh et al., 2008; Alemayehu and Lennartz, 2009). However, adsorption suffers from a major problem of saturation of bonding sites. This has limited the industrial application of physico-chemical methods (Benjamin, 1983; Mandi et al. 1993).

Nowadays, among the available metal removal processes, application of biological processes is gaining momentum due to the following reasons (Srivastava and Majumder, 2008): (1) The chemical requirements for the whole treatment process are reduced, (2) they require low operating costs, (3) they are eco-friendly and cost-effective alternative of conventional techniques, (4) they are efficient at lower levels of contamination.

Retention of heavy metals by anaerobic reactors is known to occur efficiently with > 90% removal efficiencies (La et al., 2003), and good metal removal efficiencies have therefore been achieved using Upflow Anaerobic Packed Bed reactors (UAPB) and Upflow Anaerobic Sludge Blanket reactors (UASB) as reported by many researchers (Jong and Parry, 2003, 2006; Lens et al., 2002; Van Hullebusch et al., 2006, 2007; Lenz et al., 2008). Precipitation of metals, e.g. as sulphides or carbonates, is the main removal mechanism. The reduction of sulphate to sulphides by sulfate reducing bacteria dramatically enhances the heavy metal removal efficiency compared to biosorption mechanisms due to the very low solubility of metal sulphides (Jong and Parry, 2006; La et al., 2003; Quan et al., 2003).

Different packing materials have been used for attached growth of sulphate-reducing bacteria in UAPB reactors. Dermou et al. (2006) used plastic and calcitic gravel for Cr removal. Gravel has been used by Lesage et al. (2007) to remove Co, Ni and Zn in subsurface flow constructed wetland microcosms. Plastic and rock are common packing materials for trickling filters (Tchobanoglous et al., 2003). However, the pore size and porosity of packing material play a more significant role than specific surface area in the performance of UAPB reactors (Joo-Hwa and Kuan-Yeow, 1999).

In this study, a porous volcanic rock recently investigated as alternative adsorbent for heavy metal (HM) removal (Sekomo et al., 2011) was used as packing material in a UAPB reactor because this material offers sufficient attachment surface for bacterial growth, it is cheap when locally available and it displays low bulk density compared to other rock materials. The study aims at designing a sustainable low-cost heavy metal treatment technology for textile wastewater based on a two-step process: (1) HM precipitation by sulphate reduction in anaerobic bioreactors packed with porous volcanic rocks, which is anticipated to be the major removal step with over 90 % removal; and (2) a polishing step consisting of a water hyacinth pond, for further removal of heavy metals by phytoremediation.

6.2. Materials and methods

6.2.1. Laboratory set up

The experimental set up (Figure 6-1) consisted of two different treatment compartments. The first stage was removing HM by adsorption onto the volcanic rock and by precipitation in a sulphate reducing anaerobic bioreactors. The second stage was a polishing step by means of water hyacinth plants in a pond. This polishing step was operational only during periods IV and V, after the plants had adapted to their new environment in the laboratory pond system.

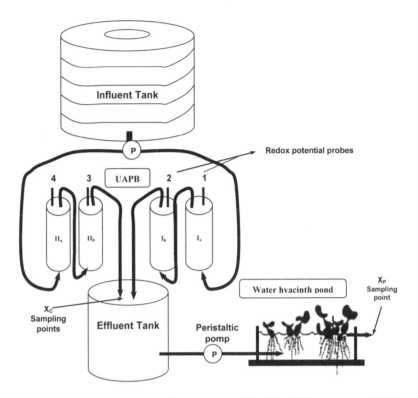

Figure 6-1: Laboratory integrated system setup, where X_c and X_p represent outlet from the columns and water hyacinth pond respectively.

The column reactors were constructed from the polyvinyl chloride (PVC) pipe with overall height of 1000 mm and 140 mm of internal diameter. Two parallel set up consisting each of 2 columns in series were used. Columns were connected by plastic tubing of 10 mm diameter. Each column was packed with porous volcanic rock collected in Gisenyi, northern region of Rwanda whose size varied between 25 – 27.5 mm. The rock has the following physical properties: 1.25 g/cm^3 of bulk density, 3.76 g/cm^3 of particle density, 2.40 $m_{eq}/100g$ of cation exchange capacity, point of zero charge at pH 7.2, a porosity of 66.7 % and 3.0 m^2/g of surface area (Sekomo et al., 2012a). A redox potential probe was placed on top of each column to monitor the potential within each column. The influent flow rate was controlled by a peristaltic pump such that the hydraulic retention time (HRT) was 1 day. Table 6-1 shows the design criteria of the UAPB.

The reactors were inoculated with sludge samples collected at KARUBANDA central prison from a sewer canal. A volume of 500 mL of sludge was added to each column before the system started to operate with sulfate reducing bacteria growth medium as synthetic wastewater of the following composition in g/L: 0.5 K_2HPO_4; 0.2 NH_4Cl; 2 $MgSO_4 \times 7H_2O$; 0.05 $FeSO_4 \times 7H_2O$; 2.5 $CaCl_2 \times 2H_2O$; 3 sodium lactate and 1 sodium citrate$\times 2H_2O$ (Jong and Parry, 2003). The temperature in the laboratory was 20 ± 3 °C. The inoculation period was conducted for 45 days in order to allow the development of a population of sulphate reducing bacteria.

Table 6-1: Design criteria of UAPB and pond system

	Design criteria	Unit	Value
PVC Filter Column	Inner diameter	cm	14
	Length	m	1
	Volume	L	≈ 11.4
	Number of columns	-	2 in series
	Flow rate	ml / min	31.16
	HRT	day	1
Packing material	Volcanic rock dimensions	mm	25 - 27.5
	Mass	kg	16.2
Pond system	Volume	L	30
	Flow rate	ml / min	20.83
	HRT	day	1

6.2.2. Biorector operation

During the experimental period, electro-conductivity (EC) and pH were directly measured in the effluents coming from the columns and the pond system using a WTW cond. 340i and a pH 340i meter kit. As mentioned before, the redox potential was measured inside the columns using a Quality in Sensing (QIS) electrode QR400X epoxy redox-electrode with platinum as indicator electrode and silver / silver chloride reference electrode. The system was operated under different concentration regimes: Periods I and III were conducted with 5 mg/L of each heavy metal; Periods II and V with 10 mg/L of each heavy metal and Period IV with concentrations approximating those found in real wastewater from the textile industry (0.5 mg/L for Cd, 1 mg/L for Cu and Pb and 5 mg/L for Zn) as indicated on page 9. Heavy metal solutions were prepared by dissolving the appropriate weight of the respective salts $Cd(NO_3)_2 \times 4H_2O$, $Cu(NO_3)_2 \times 2H_2O$, $Pb(NO_3)_2$, $Zn(NO_3)_2 \times 6H_2O$ in tap water. After each period, a resting time of two days was taken, the time to clean the influent tank and prepare wastewater for another run. During that time all inlet valves from the influent tank were closed in order to avoid emptying the columns and keep the bacterial activity within the columns.

6.2.3. Sampling

Water samples were collected daily at the outlet of the columns and of the water hyacinth pond (Figure 6-1) for analysis of dissolved and total heavy metal concentrations. After 130 days of operation, the system was stopped and sludge samples were collected at the bottom of each column and in the effluent tank. Water hyacinth plants were collected from the pond system and divided into three parts (root, stem and leaves) to assess metal accumulation and uptake in the plants.

6.2.4. Analytical procedures

After collection of water samples and filtration through a Whatman filter paper of 45 μm, a few drops of HNO_3 65 % were added to the samples immediately prior to dissolved heavy metal analysis. Plants and sludge samples were first dried for 24 hours at 103 (± 2) °C. Subsamples were then digested during at least 7 hours using a mixture of concentrated HCl, HNO$_3$ and H$_2$O$_2$. The digestion temperature was varied from 100 °C to 200 °C, in steps of 25 °C each 30 minutes until the maximum temperature was reached. After the digestion step, the mixture was cooled and transferred to a 250 mL clean volumetric flask, filled up to the mark with distilled water, mixed and then allowed to settle for at least 15 hours (Kansiime et al., 2007).

All samples were then analysed on a Flame Atomic Absorption Spectrometer (Perkin Elmer model AAnalyst 200). The analysis was done in duplicate. A mass balance analysis for heavy metal within the column system has been established based on the following equation:

Influent = Effluent (eff. soluble + eff. particulate) + Sludge + Adsorption / Biosorption (1)

In this equation, heavy metals are being removed from the system by sulphide precipitation and adsorption combined with biosorption onto porous volcanic rock. As the effluent solution contained suspended particulates, at the end of each running period the sludge from the effluent tank was analysed and giving the effluent particulate fraction in the mass balance expression. Furthermore, as the sulphide precipitation is known to be the major and long term removal mechanism for heavy metals, it was assumed that the difference in metal concentration in the above relation between the influent concentration at one hand and the effluent plus sludge concentrations at the other hand will represent the fraction of heavy metal being adsorbed on the adsorbent and its associated biofilm. COD was determined by a titrimetric method using a mixture of chromic and sulphuric acids to oxidize organic matter. Sulphate ions were determined by the turbidimetric method where sulphate ions were precipitated in an acetic acid medium with BaCl$_2$ and the light absorbance measured by a photometer (APHA, 1992).

6.2.5. Data analysis

Statistical analysis was performed with Excel Stat™, in order to assess differences in COD, in heavy metal and sulfate removal between columns I and II. A paired-two sample for means (T-Test) was used to assess differences between mean values of low and high concentration solutions respectively, 5 mg/L and 10 mg/L of heavy metal. The two-sample assuming unequal variances (T-Test) was also used to assess differences between columns I and II during different operational periods ($\alpha = 0.05$).

6.3. Results

6.3.1. Heavy metal removal in the columns

All heavy metals were efficiently removed from the influent. In general, the removal efficiency varied between 90 – 97 % for Cd, 96 – 100 % for Cu, 84 – 99 % for Pb and 63 – 97 % Zn (Figure 6-2). During the switching from low to high heavy metal concentrations, a decrease in the removal efficiency occurred, especially for Zn and Cd. No significant differences in heavy metal removal were observed (P > 0.05) when comparing the two parallel systems of columns.

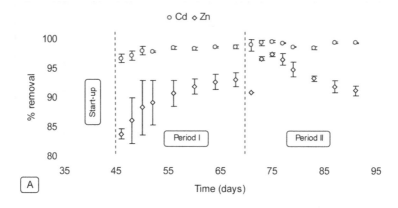

A

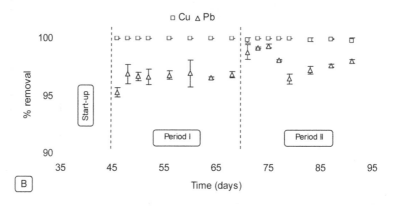

B

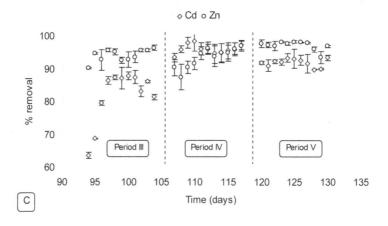

C

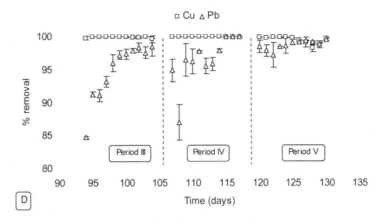

Figure 6-2: Removal efficiency in the columns for different experimental periods A, B, C and D [Low strength (period I & III), high strength (period II & V) and real onsite (period IV) n = 4 heavy metal concentrations].

After a running period of 130 days, the sludge from the reactor and the effluent tank was collected for heavy metal analysis. In general, high metal concentrations were recorded in the sludge from the inlet columns (Columns I_A and II_A), whereas low concentrations were recorded in the outlet columns (Columns I_B and II_B) and in the effluent tank (Figure 6-3). The mass balance analysis showed that heavy metals were mainly accumulated in the sludge (Table 6-2).

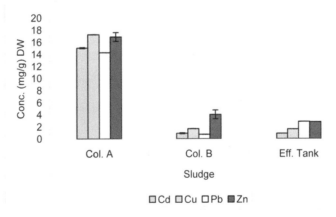

Figure 6-3: Heavy metal concentration in the sludge taken from the columns (n = 2) and effluent tank.

Table 6-2: Mass balance analysis and overall % removal efficiency of the column system.

Metal	Influent Solution (mg)	Effluent Solution (mg)	Effluent Particulate (mg)	Sludge (mg)	Adsorption / Biosorption (mg)	% removal
Cd	31.05	0.81	4.21	16.01	10.02	97.3
Cu	31.73	0.08	5.56	18.99	7.09	99.8
Pb	31.19	0.81	4.28	15.03	11.07	97.4
Zn	35.15	3.22	8.82	20.94	2.16	90.8

6.3.2. Variation of COD, SO_4^{2-}, EC, pH and redox potential in the columns

The reactor performance was assessed by the monitoring of several physical and chemical parameters. In general, the COD removal efficiency varied from 36 – 91 % and sulphate removal efficiency from 65 – 98 % during different running periods (Figure 6-4). No significant differences between column A and B (P > 0.05) in EC, COD, pH and sulphate ions were observed. However, significant differences between column A and B (P < 0.012) in redox potential were observed.

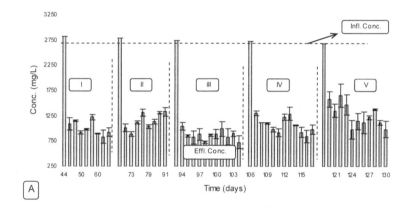

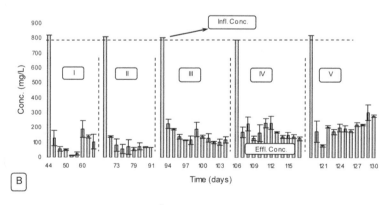

Figure 6-4: Variation of COD and SO_4^{2-} (A & B), the dashed horizontal line represents all influent concentrations and the dashed vertical line separates experimental periods respectively I, II, III, IV & V (n = 4).

During the operating period the conductivity in the effluent of the column systems continuously increased. The pH was more or less constant during the first and second period. Thereafter, it decreases during period III and V with an intermediate recovery during period IV.

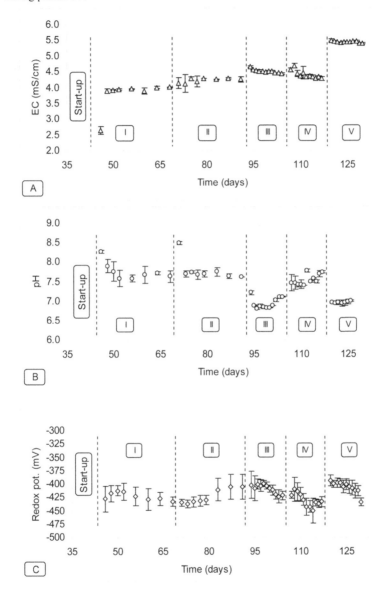

Figure 6-5: Variation of EC, pH and redox potential in the columns, (A, B and C) I, II, III, IV & V stands for periods respectively (n = 2 for EC & pH and n = 4 for Eh).

6.3.3. Heavy metal removal in the water hyacinth pond

The visual observation of water hyacinth growing in the effluent of the columns during period IV and V showed some changes in the leaves colour after one week of operation. A green-yellow colour was noticed and some partial wilting and brown leaves appeared after two weeks, but in general the plant survived until the end of the experiment (23 days). The pond system with water hyacinth removed further heavy metals from the effluent tank containing the water from the column systems. Figure 6-6 compares the inlet and outlet of the systems. The t-test revealed significant differences in Cd and Pb effluent concentration between the column and the pond system (P < 0.016); However, no significant differences (P > 0.05) were revealed for Cu and Zn. When the system started to operate at high heavy metal concentrations (period V), a change in the outlet metal concentration was noticed when compared with the previous run (period IV).

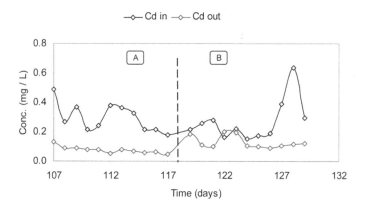

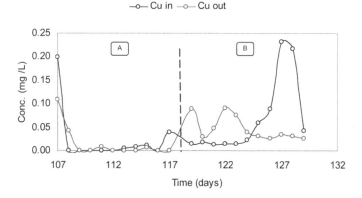

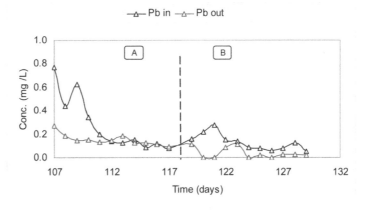

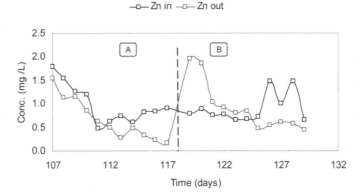

Figure 6-6: Heavy metal concentration at the columns (solid black line) and pond system (solid blue line) outlets. A and B regions represent data collected for periods IV and V, respectively.

Analysis of the water hyacinth plants after the experimental period (Figure 6-7) showed that accumulation of heavy metals on the root system was the main removal mechanism. Uptake and translocation were also noticed but seemed not as important as accumulation at the roots. Concentrations of heavy metal in different water hyacinth parts (roots, shoots and leaves) after the experiment were compared with those recorded in control plants (water hyacinth plants analyzed prior to exposure to the polluted water from the column system). This revealed significant differences ($P <$ 0.05) for all metals in the roots and for Zn in the leaves. No significant differences could be demonstrated for neither of the metals in the shoots, and for Cd, Cu and Pb in the leaves. Cu and Zn were the most accumulated and taken up by the plant. Comparison of the plant results obtained under periods IV and V showed that the longer the exposure to the pollutant or to a high pollutant concentration, the higher the metal concentration will be accumulating in the plant (Figure 6-7).

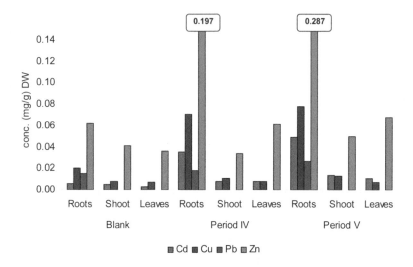

Figure 6-7: Heavy metal concentration in different parts of the water hyacinth plant grown under different initial heavy metal concentrations.

6.4. Discussion

6.4.1. Columns system

This study showed that heavy metals can be efficiently removed (> 90 %) in Upflow Anaerobic Packed Bed reactors. The heavy metal concentrations remaining in the effluent were in general below the WHO (2006) standards for irrigation except for Cd. The recommended maximum concentration for Cd is 0.01 mg/L However, for irrigation purposes a complete analysis of the wastewater is required in order to select appropriate crops that could be irrigated with such kind of water is very important as nutrients demand differ in general. Results from this study gave the following order in metal removal Cu > Pb > Cd > Zn. This is in good agreement with what was reported by Tchobanoglous et al. (2003), when comparing solubility product (pKs) values of the metal sulphides of Cd, Cu, Pb and Zn.

The mass balance analysis (Table 6-2) showed that metals were mainly removed by accumulation in the sludge (61.9 % – 84.7 % of total removal), probably as metal sulphide precipitates. Furthermore, the use of porous volcanic rock played an additional role in metal removal by adsorption combined with biosorption, contributing with 6.2 % – 32.3 % to the overall removal. This affinity for heavy metals has been reported before and for this particular type of volcanic rock resulted in maximum sorption capacities of 6.29 mg/g for Cd, 10.87 mg/g for Cu, 9.52 mg/g for Pb and 4.46 mg/g for Zn (Sekomo et al., 2011). Table 2 shows that these maximum sorption capacities have not been exceeded mainly because of the presence of other competing metal removal mechanisms like biosorption.

The COD/SO_4^{2-} ratio of a wastewater is an important parameter controlling the electron flow in anaerobic biorectors. When that ratio is higher than 6, methanogenic bacteria will be predominant, but at a ratio less than 6, sulphate reducing bacteria (SRB) will be more competitive, producing more H_2S that affects the methanogenic bacteria but not the SRB (Mizuno et al., 1994). According to other researchers (Lens et al., 1998; Weijma et al., 2002; Thabet et al., 2009), lowering further the ratio COD/SO_4^{2-} promoted sulphate reduction over methanogenesis. In this study, the ratio COD/SO_4^{2-} was 3, thereby favouring SRB over methanogenic bacteria.

The average COD removal in the columns varied between 52.3 % and 68.3 % whereas the average SO_4^{2-} removal varied between 75.1 % and 90.5 % for all the periods (data not shown). When comparing the performance of the system in terms of COD and SO_4^{2-} removal for the entire experiment, it appears that there was in general a decreasing pattern in the system performance. This could be due to the effect of increasing heavy metal concentrations with a potential toxic effect of heavy metals on the SRB. Cabrera et al. (2006) reported on the toxic effects of dissolved heavy metals on SRB, and concluded that the SRB showed different responses to different metals and different concentration. This was further supported by Garcia et al. (2001) who reported a decrease in sulphate reduction at concentrations of Cu and Fe above 100 mg/L and 30 mg/L. Medircio et al. (2006) observed a 30 % decrease in sulphide removal due to Cd toxicity at a concentration of 19.5 mg/L. In our study, the Cd concentration was 2 to 3 times less than the concentration used by Medircio et al. (2006) A decrease of 14.1 % in sulphide and 12.1 % in COD removal efficiency was observed. Our results show that even at low concentrations, the system performance is affected to some extent. Despite the decrease in COD and sulphate removal efficiency, the overall metal removal efficiency was not affected. This is backed up by the mass balance analysis (Figure 6-4), which shows that the sulphate concentration decreased from an average of 811 mg/L to 137 mg/L, corresponding to 674 mg/L of sulfate consumed and 225 mg/L sulphide produced. Taking into account the stochiometry (1 mole of sulphide, after its production, would eliminate 1 mole of metal), this amount of sulphide is sufficient to precipitate a mixture of 788 mg/L of Cd, Cu, Pb and Zn.

Bacterial activities in the column can decrease or increase the conductivity of the solution as a result of the production of charged mobile metabolites such as organic acids and the decomposition of macromolecules into simple compounds (Jong and Parry, 2006). The conductivity measured in the effluent varied depending on the experimental period (Figure 6-5A), with in general an increasing EC from period I to V. During period II and V, the influent heavy metal concentration was increased to 10 mg/L. This slight increase and stabilization in EC recorded in each period is an indication that anaerobic bacteria were accommodated to the sulphate-rich environment.

The speciation of sulphide in water depends on the pH value where two major species do exist in aqueous solution (H_2S and HS^-). Since the pH in the column effluent was around 7.4 during the experimental period, the major fraction of sulphide produced was in dissolved form (HS^-), responsible for metal precipitation. This value was close to the interval of pH reported by Silva et al. (2002) in an anaerobic reactor treating sulphate-rich industrial water which varied between 7.9 and 9.2; where 0.2 % to 12.1 % of total sulphide produced was found as gaseous H_2S.

Figure 6-5B illustrates effluent pH measurements at the column outlets. The average pH was 6.7 ($\pm$ 0.3) in the influent, it increases to 7.4 ($\pm$ 0.1) in the effluent due to lactate degradation by SRB occurring in the reactor to produce hydrogen sulphide and bicarbonate, as shown in the following reactions:

$$2\ CH_2O + SO_4^{2-} \rightarrow S^{2-} + 2\ CO_2 + 2\ H_2O \quad (2)$$
$$S^{2-} + 2\ CO_2 + 2\ H_2O \rightarrow H_2S + 2\ HCO_3^- \quad (3)$$

The bicarbonate produced increases the pH of the solution. Most reduction reactions consume H^+, resulting for instance in a pH rise in acid soils (Patra & Mohanty 1994, Narteh & Sahrawat 1999). It is CO_2 in the reduced environment that buffers pH via the H_2CO_3 / HCO_3^- reaction (McBride, 1994; Elliot et al., 1998).

The average inlet redox potential was +131.2 ($\pm$ 18.2) mV but decreases to -419.0 ($\pm$ 12.7) mV (Figure 6-5C) in the effluent as a result of the SRB activities on the wastewater. This decrease in Eh is due to the O_2 consumption by microbial activities and H_2S production. Sulphate reduction is initiated at a Eh below $-$ 100 mV (Reddy and Graetz 1988). The level of redox reduction is higher in the presence of organic matter because organic matter is oxidized and soil components are reduced by anaerobic microbial respiration (Ponnamperuma 1972).

6.4.2. Water hyacinth pond system

From this study, the pond system as potential polishing step for the effluent of the UAPB columns was investigated. This step further removed heavy metals by phytoremediation (Figure 6-7) and is thus an important step in the overall treatment process. The concentrations recorded in the wastewater at the pond outlet showed low HM concentrations when compared to the concentration recorded at the column outlets. This indicates that the water hyacinth was accumulating heavy metals from the polluted water. Many other researchers have reported the potential of aquatic macrophytes in metal removal from polluted water (Deng et al., 2004; Miretzky et al., 2004; Liu et al., 2007; Mishra and Tripathi, 2009). The variation in the concentration of the pond effluent was due to changing metal concentrations from low to high in the columns, as shown by the dashed line between region A and B (Figure 6-6).

Soltan and Rached (2003) reported on different toxicity effects of two heavy metals known to be very toxic to plants (Cd and Pb) with a mixture of other heavy metal (Co, Cr, Cu, Mn, Ni and Zn) on water hyacinth plants. Their results showed that water hyacinth could survive in a mixture of heavy metals with concentrations up to 3 mg/L for each heavy metal and 100 mg/L of Pb, whereas higher concentrations of metals as mixtures and 100 mg/L of Cd led to rapid fading of the plants. In our study, heavy metal concentrations were five to ten times less. From the visual changes observed on plants growing on the UAPB effluent, it can be inferred that the toxicity of the effluent for the plants increases with the increasing contact time with the polluted water. Therefore, the toxicity observed is probably not only due to heavy metals but also to the excess of sulphide contained in the water.

For some aquatic macrophytes such as *Stratiotes aloides* and *Potamogeton compressus*, sulphide concentrations as low as 3.4 µg/L can be toxic (Smolders and Roelofs, 1995). Chambers et al. (2005) reported that high sulphide concentrations were negatively affecting *Phragmites,* but possibly also high chloride concentrations (11.59 g/L). In our study, the sulphide concentration was more than 1,5 times the toxic concentration reported by Hotes et al. (2005) for *Phragmites,* so this could also explain the wilting of the water hyacinth leaves.

The plant contribution to heavy metal removal was showed by the analysis of different parts of the plant (roots, shoot and leaves) before and after exposure to the polluted water (Figure 6-7). Results from this study showed mainly removal of heavy metal via accumulation on the roots; however translocation was also noticed. The accumulation at the root occurred in the following order: Zn > Cu > Cd > Pb. In many investigations conducted on aquatic plants, it was observed that heavy metals have been mainly accumulated on the roots (William et al., 2000; Aksoy et al. 2005; Amri et al. 2007 and Sekomo et al. 2010). Furthermore, Cu and Zn were accumulated to a larger extent then Cd and Pb. This can be due to the fact that Cu and Zn, like Fe, are essential micronutrients for plant metabolism; but when present in excess, they become toxic to the plant (William et al., 2000).

The integrated system performance was good in heavy metal removal. In order to lower further the outlet heavy metal concentration, it would be possible to add two or more ponds in series to the first one and then the system could reach very good removal efficiencies (Rousseau et al., 2011). Furthermore in order to avoid the sulphide toxicity to the plant, addition of lime ($Ca(OH)_2$) can be undertaken to raise the pH above 8 and shift the H_2S equilibrium towards the non volatile HS^- in the settling tank before the wastewater can be treated by the plants. This action can also result in the lowering of heavy metal concentrations due to the formation of metal hydroxide precipitates. Judith and Peddrick (2004) stressed the importance of knowing the processes of metal removal, uptake and distribution in the wetland. The extent of uptake and how metals are distributed within plants can have important effects on the residence time of metals in plants, in wetlands and the potential release of metals when conditions change.

6.5. Conclusions

UAPB reactors were characterized by a high metal removal efficiency of over 90 % in the following order: Cu > Pb > Cd > Zn. Simultaneous removal of COD and sulphate was also achieved. The mass balance analysis showed that sulphide precipitation was the major removal mechanism followed by adsorption and biosorption in the column. Switching from low to high strength metal concentration did not affect much the performance of the reactor. Coupling a water hyacinth pond to the system showed a further improvement in the metal removal efficiency, with heavy metals mainly accumulated to the root system. Additional measures need, however, to be taken to avoid sulphide toxicity to the plants. Lime addition can be undertaken to raise the pH above 8 and shift the H_2S equilibrium towards the non volatile HS^- in the settling tank before the wastewater can be further treated by the plants. Therefore, the integrated system can be recommended as alternative low cost and sustainable technology for heavy metal removal for developing countries.

Acknowledgements

The authors are grateful to the NUFFIC / Dutch Government and the National University of Rwanda for the financial support provided for this research through the Netherlands Programme for Institutional Strengthening of Post-secondary Education and Training Capacity (NPT / RWA / 051-WREM) Project. The authors would like also to thank the laboratory staff for UNESCO-IHE for their technical assistance during this research.

6.6. References

Ahmet, S., Mustafa, T. & Mustafa, S. (2007). Adsorption of Pb(II) and Cr(III) from aqueous solution on Celtek clay. Journal of Hazardous Materials 144, 41 – 46.

Aksoy, A., Demirezen, D. & Duman, F. (2005). Bioaccumulation, detection and analyses of heavy metal pollution in Sultan Marsh and its environment. Water Air and Soil Pollution 164, 241 – 255.

Alemayehu, E. & Lennartz, B. (2009). Virgin volcanic rocks: Kinetics and equilibrium studies for the adsorption of cadmium from water. Journal of Hazardous Materials 169, 395 – 401.

Ali, N., Hameed, A. & Ahmed, S. (2009). Physicochemical characterization and Bioremediation perspective of textile effluent, dyes and metals by indigenous Bacteria. Journal of Hazardous Materials 164, 322 – 328.

American Public Health Organisation, (1992). Standards Methods for examination of Water and Wastewater 18th edition.

Amri, N., Benslimane, M., Zaoui, H., Hamedoun, M. & Outiti, B. (2007). Evaluation Of The Heavy Metals Accumulate In Samples Of The Sediments, Soils And Plants By ICPOES With The Average Sebou. M. J. Condensed Matter. 8 (1), 43 – 52.

Banat, I.M., Nigam, P., Singh, D. & Marchant, R. (1996). Microbial decolorization of textile-dye-containing effluents: a review. Bioresource Technology 58, 217 – 227.

Benjamin, M.M. (1983). Adsorption and surface precipitation of metals on amorphous iron oxyhydroxide. Environmental Science and Technology 17, 686 – 692.

Beszedits, S. (1988). Chromium removal from industrial wastewater, in: O. Nriagu, E. Nieboer (Eds.), Chromium in the Natural and Human Environments, John Wiley, New York.

Cabrera, G., Pérez, R., Gomez, J.M., Abalos, A. & Cantero, D. (2006). Toxic effects of dissolved heavy metals on Desulfovibrio vulgaris and Desulfovibrio sp. Strains. Journal of Hazardous Materials 135 (1-3), 40 – 46.

Chambers, R.M., Mozdzer, T.J. & Ambrose, J.C. (1998). Effects of salinity and sulphide on the distribution of Phragmites australis and Spartina alterniflora in a tidal marsh. Aquatic Botany 62, 161 – 169.

Daza, L., Mendioroz, S. & Mendioroz, J.A. (1991). Mercury adsorption by sulfurized fibrous silicates, Clays Clay Minerals 39, 14 – 21.

Deng, H., Ye, Z.H. & Wong, M.H. (2004). Accumulation of lead, zinc, Chromium and Zinc by 12 wetland plants species thriving in metal contaminated sites in China, Environmental Pollution 132, 29 – 40.

Dermou, E., Velissariou, A., Xenos, D. & Vayenas, D.V. (2007). Biological removal of hexavalent chromium in trickling filters operating with different filter media types. Desalination, 211 (1-3), 156 – 163.

Elliott, P., Ragusa, S. & Catcheside, D. (1998). Growth of sulphate-reducing bacteria under acidic conditions in an upflow anaerobic bioreactor as a treatment system for acid mine drainage. Water Research 32 (12), 3724 – 3730.

Erdem, E., Karapinar, N. & Donat, R. (2004). The removal of heavy metal cations by natural zeolites. Journal of Colloid and Interface Science 280, 309 – 314.

Fu, Y. & Viraraghavan, T. (2001). Fungal decolorization of dye wastewater: A review. Bioresource Technology 79, 251 – 262.

Garcia, C., Moreno, D.A., Ballester, A., Blazquez, M.L. & Gonzalez, F. (2001). Bioremediation of an industrial acid mine water by metal-tolerant sulphate reducing bacteria. Minerals Engineering 14 (9), 997 – 1008.

Hotes, S., Adema, E.B., Grootjans, A.P., Takashi, I. & Poschlod, P. (2005). Reed die-back related to increased sulphide concentration in acoastal mire in eastern Hokkaido, Japan. Wetland Ecology and Management 13, 83 – 91.

Jong, T. & Parry, D.L. (2003). Removal of sulphate and heavy metals by sulphate reducing bacteria in short-term bench scale upflow anaerobic packed bed reactor runs. Water Research 27, 3379 – 3389.

Jong, T. & Parry, D.L. (2006). Microbial sulphate reduction under sequentially acidic conditions in an upflow anaerobic packed bed bioreactor, Water Research 40 (13), 2561 – 2571.

Joo-Hwa, T., Jeyaseelan, S. & Kuan-Yeow, S. (1999). Performance of anaerobic packed bed system with different media characteristics. Water Science and Technology 34 (5-6), 453 – 459.

Juang, R.S., Tseng, R.L., Wu, F.C. & Lin, S.J. (1996). Use of chitin and chitosan in lobster shell wastes for colour removal from aqueous solutions. Journal of Environmental Science and Health A. 31, 325 – 338.

Judith, S.W. & Peddrick, W. (2004). Metal uptake, transport and release by wetland plants: implications for phytoremediation and restoration. Environment International 30, 685 – 700.

Kansiime, F., Saunders, M.J. & Loiselle, S.A. (2007). Functioning and dynamics of wetland vegetation of Lake Victoria: an overview. Wetland Ecology and Management 15, 443 – 451.

La, H.J., Kim, K.H., Quan, Z.X., Cho, Y.G. & Lee, S.T. (2003). Enhancement of sulphate reduction activity using granular sludge in anaerobic treatment of acid mine drainage. Biotechnology Letters 25, 503 – 508.

Lens, P.N.L., Vallero, M., Esposito, G. & Zandvoort, M. (2002). Perspectives of sulphate reducing bioreactors in environmental biotechnology. Reviews in Environmental Science and Biotechnology 1(4), 311 – 325.

Lens, P.N.L., Visser, A.N.L., Janssen, A.J.H., Hulshoff Pol, L.W. & Lettinga, G.W. (1998). Biotechnological treatment of sulphate-rich wastewaters, Critical Reviews in Environmental Science and Technology 28, 41 – 88.

Lenz, M., Van Hullebusch, E.D., Hommes, G., Corvini, P.F.X. & Lens, P.N.L. (2008). Selenate removal in methanogenic and sulphate-reducing upflow anaerobic sludge bed reactors. Water Research 42, 2184 – 2194.

Lesage, E., Rousseau, D.P.L., Van de Moortel, A., Tack, F.M.G., De Pauw, N. & Verloo, M.G. (2007). Effects of sorption, sulphate reduction, and Phragmites australis on the removal of heavy metals in subsurface flow constructed wetland microcosms. Water Science and Technology. 56 (3), 193 – 198.

Liu, J., Donga, Y., Xu, H., Wang, D. & Xu, J. (2007). Accumulation of Cd, Pb and Zn by 19 wetland plant species in constructed wetland, Journal of Hazardous Materials 147, 947 – 953.

Mandi, L., Houhowm, B., Asmama, S. & Schwartzbrod, J. (1996). Wastewater treatment by reed beds: an experimental approach. Water Research 30 (9), 2009 – 2016.

Marcucci, M., Ciardelli, G., Matteucci, A., Ranieri, L. & Russo, M. (2002). Experimental campaigns on textile wastewater for reuse by means of different membrane processes. Desalination 149, 137 – 143.

McBride, M.B. (1994). Environmental chemistry of soils. Oxford. University Press, New York, NY.

Medircio, S.N., Leao, V.A. & Teixeira, M.C. (2007). Specific growth rate of sulphate reducing bacteria in the presence of manganese and cadmium. Journal of Hazardous Materials 143 (1-2), 593 – 596.

Miretzky, P., Saralegui, A. & Cirelli, F. (2004). Aquatic macrophytes potential for the simultaneous removal of heavy metals (Buenos Aires, Argentina), Chemosphere 57, 997 – 1005.

Mishra, V.K. & Tripathi, B.D. (2009). Accumulation of chromium and zinc from aqueous solutions using water hyacinth (Eichhornia crassipes). Journal of Hazardous Materials 164, 1059 – 1063

Mizuno, O., Li, Y.Y. & Noike, T. (1994). Effect of sulphate concentration and sludge retention time on the interaction between methane production and sulphate reduction for butyrate, Water Science and Technology 30, 45 – 54.

Murari, P., Huan-yan, X. & Sona, S. (2008). Multi-component sorption of Pb(II), Cu(II) and Zn(II) onto low-cost mineral adsorbent. Journal of Hazardous Materials 154, 221 – 229.

Narteh, L.T. & Sahrawat, K.L. (1999). Influence of flooding on electrochemical and chemical properties of West African soils. Geoderma 87, 179 – 207.

Ozdemira, O., Turana, M., Zahid, T.A., Fakia, A. & Baki, E.A. (2009). Feasibility analysis of color removal from textile dyeing wastewater in a fixed-bed column system by surfactant-modified zeolite (SMZ). Journal of Hazardous Materials 166, 647 – 654.

Pansini, M., Colella, C. & Gennaro, M.D. (1991). Chromium removal from wastewater by ion-exchange using Zeolite. Desalination 83 (1-3), 145 – 157.

Patra, B.N. & Mohanty, S.K. (1994). Effects of nutrients and liming on changes in pH, redox potential, and uptake of iron and manganese by wetland rice in iron-toxic soil. Biology and Fertility of Soils 219 (17), 285 – 288.

Perez-Candela, M., Martin-Martinez, J.M. & Torregrosa-Macia, R. (1995). Chromium(VI) removal with activated carbon. Water Research 29 (9), 2174 – 2180.

Ponnamperuna, F.N. (1972). The chemistry of submerged soils. Advances in Agronomy 24, 29 – 96.

Quan, Z.X., La, H.J., Cho, Y.G., Hwang, M.H., Kim, L.S. & Lee, S.T. (2003). Treatment of metal contaminated water and vertical distribution of metal precipitates in an upflow anaerobic bioreactor. Environmental Technology 24, 369 – 376.

Reddy, K.R. & Graetz, D.A. (1988). The influence of redox potential on the environmental chemistry of contaminants in soils and sediments. In: Hook D. (ed.), The Ecology and Management of Wetlands. Timber Press.

Rengaraj, S., Kyeong-Ho, Y. & Seung-Hyeon, M. (2001). Removal of chromium from water and wastewater by ion-exchange resins. Journal of Hazardous Materials 87, 273 – 287.

Rousseau, D.P.L., Sekomo, C.B., Saleh, S.A.A.E. & Lens, P.N.L. (2011). Duckweed and Algae Ponds as a Post-Treatment for Metal Removal from Textile Wastewater. In: Vymazal J. (Ed), Water and Nutrient Management in Natural and Constructed Wetlands, Springer Science, Dordrecht.

Sekomo, C.B., Nkuranga, E., Rousseau, D.P.L. & Lens, P.N.L. (2011). Fate of Heavy Metals in an Urban Natural Wetland: The Nyabugogo Swamp (Rwanda). Water Air and Soil Pollution 214, 321 – 333.

Sekomo, C.B., Rousseau, D.P.L. & Lens, P.N.L. (2011). Adsorptive removal of Cd(II), Cu(II), Pb(II) and Zn(II) from the aqueous phase using volcanic rock as adsorbent. Water Air and Soil Pollution., DOI 10.1007/s11270-011-0880-z

Silva, A.J., Varesche, M.B., Foresti, E. & Zaiat, M. (2002). Sulphate removal from industrial wastewater using a packed-bed anaerobic reactor. Process in Biochemistry 37, 927 – 935.

Smolders, A. &. Roelofs, J.G.M. (1995). Sulphate mediated iron limitation and eutrophication in aquatic ecosystems. Aquatic Botany 45, 247 – 253.

Soltan, M.E. & Rashed, M.N. (2003). Laboratory study on the survival of water hyacinth under several conditions of heavy metal concentrations. Advances in Environmental Research 7, 321 – 334.

Srivastava, N.K. & Majumder, C.B. (2008). Novel biofiltration methods for the treatment of heavy metals from industrial wastewater: A review. Journal of Hazardous Materials 151, 1 – 8.

Suh, C.E., Luhr, J.F. & Njome, M.S. (2008). Olivine-hosted glass inclusions from Scoriae erupted in 1954 – 2000 at Mount Cameroon volcano, West Africa. J. Volcanol. Geotherm. Res. 169 (1–2), 1 – 33.

Tchobanoglous, G., Burton, F.L. & Stensel, H.D. (2003). Waste Water Engineering: Treatment and Reuse, 4th ed., McGraw-Hill Higher Education, New York.

Thabet, O.B.D., Bouallagui, H., Cayol, J.L., Olivier, B., Fardeau, M.L. & Hamdi, M. (2009). Anaerobic degradation of landfill leachate using an upflow anaerobic fixed-bed reactor with microbial sulphate reduction. Journal of Hazardous Materials 167, 1133 – 1140

Van Hullebusch, E.D., Farges, F., Lenz, M., Lens, P.N.L. & Brown, G.E.J. (2007). Selenium speciation in biofilms from granular sludge bed reactors used for wastewater treatment. Am. Inst. Phys. Conf. Proc. 882, 229 – 231.

Van Hullebusch, E.D., Gieteling, J., Zhang, M., Zandvoort, M.H., Daele, W.V., Defrancq, J. & Lens, P.N.L. (2006). Cobalt sorption onto anaerobic granular sludge: isotherm and spatial localization analysis. Journal of Biotechnology 121 (2), 227–240.

Weijma, J., Bots, E.A.A., Tandlinger, G., Stams, A.J.M., Hulshoff Pol, L.W. & Lettinga, G.W. (2002). Optimisation of sulphate reduction in amethanol-fed thermophilic bioreactor, Water Research 36, 1825 – 1833.

WHO (2006). Guidelines for the safe use of wastewater, excreta and grey water. Volume 2 wastewater use in agriculture.

Williams, L.E., Pittman, J.K. & Hall, J.L. (2000). Emerging mechanisms for heavy metal transport in plants. Biochimica et Biophysica Acta 1465 (12), 104 – 126.

Chapter 7: General discussion and conclusions

7.1. Heavy metal removal in natural wetlands

Wetlands are natural purifiers for water in the environment through the ages. However, the economic development in the world is putting considerable pressure on these wetlands with dumping of all kind of pollutants, therefore threatening the entire equilibrium of the ecosystem. This study investigated the heavy metal concentrations flowing through an urban natural wetland case, the Nyabugogo wetland in Rwanda. The metal concentration in water flowing through the Nyabugogo swamp has been determined in water, in *C. papyrus*, in sediment, in fish (*Clarias sp.* and *Oreochromis sp.*), and in *Oligochaetes*. Although the results showed that there is a considerable metal accumulation in plants and in sediments, the study showed that there is also a high concentration of heavy metals in the water body of the swamp and in the outflow. There is in general a pollution problem caused by Cd and Pb in the water phase. The highest accumulation of heavy metals has been found both in the plants and the sediments. Roots play a more dominant role in heavy metal removal compared to the stem and the umbel. Fish in the Nyabugogo swamp showed a high concentration of heavy metals, especially Cr, Cd, and Pb. In general, there is a genuine concern for human health through direct consumption and uptake of metals via drinking water and eating fish, but also indirectly by using the polluted water for irrigation of crops around the wetland.

Wetlands are among the worlds most productive ecosystems. Aquatic macrophytes such as *C. papyrus, Eichhornia crassipes* and *Phragmites spp.* exhibit high growth rates and net primary production, compared to agricultural crops like maize and sugar cane (Kansiime et al., 2007). However, as the receiving body, the wetland retains all kinds of pollutants from the wastewaters. Heavy metals are a great concern due to their toxicity to aquatic life and human health already at trace levels (Wang et al., 2006). As reported by Williams et al. (2000) and Manios et al. (2003), there is a threshold of tolerance of each plant to heavy metal accumulation. For a number of environmental, physiological, and genetical reasons this threshold is different among plant species. When this limit is passed, the toxic effect of metals in plants takes place, and metals become poisonous. Results of this study are above that limit, showing that the Nyabugogo swamp is a metal polluted environment. The heavy metal concentrations recorded in the Nyabugogo swamp were in general below the toxic limit for Cu, Pb and Zn. However, some isolate high concentrations were recorded for Cu and Zn. *C. papyrus* accumulated, took up, and translocated more Zn when compared to the other heavy metals as reported by Sekomo et al. (2011a). This could somehow explain the situation occurring in crops cultivated by people in the swamp using the polluted water for crop irrigation.

Heavy metal pollution in the flesh of fishes was in general above the limits. This study shows that the Cr content in *Clarias sp.* and *Oreochromis sp.* is high and presents a risk (Denny, 1995). Cd and Pb values are also high in fish when compared to the limits fixed by the European Commission on heavy metals in fish (2001). Food ingestion is normally a more important source of metals contamination than drinking water. Therefore, the accumulation of metals in tissue can result from eating habits of fishes (Monday et al. 2003). It is clear how the ingestion of metal contaminated *Oligochaetes* and the direct contact with contaminated water might be the principal routes of heavy metals exposure to fish in wetlands. Heavy metal concentrations from the present study are higher than the standard limit and thus present a human health concern for people consuming fish from the swamp. In general, there is a genuine concern for human health through direct consumption and uptake of metals via drinking water and eating fish, but also indirectly by using the polluted water for irrigation of crops around the wetland. The wetlands of the Lake Victoria basin are characterized by highly productive emergent macrophytic plant communities (Jones and Humphries 2002) where *C. papyrus L.* forms the monotypic vegetation in permanently flooded areas. Tropical emergent macrophytes have been shown to exhibit high rates of the aboveground net productivity.

Unlike organic contaminants, heavy metals are not biodegradable and tend to accumulate in living organisms and many heavy metal ions are known to be toxic or carcinogenic. Toxic heavy metals of particular concern in the treatment of industrial wastewaters include zinc, copper, cadmium, lead and chromium (Fu and Wang, 2011). Zinc is a trace element essential for human health. It is important for the physiological functions of living tissue and regulates many biochemical processes. However, too much zinc can cause eminent health problems, such as stomach cramps, skin irritations, vomiting, nausea and anemia (Oyaro et al., 2007). Copper does essential work in animal metabolism. But the excessive ingestion of copper brings about serious toxicological concerns, such as vomiting, cramps, convulsions, or even death (Paulino et al., 2006). Cadmium has been classified by U.S. Environmental Protection Agency as a probable human carcinogen. Cadmium exposes human health to severe risks. Chronic exposure of cadmium results in kidney disfunctioning and high levels of exposure will result in death. Lead can cause central nervous system damage. Lead can also damage the kidney, liver and reproductive system, basic cellular processes and brain functions. The toxic symptoms are anemia, hallucination, headache, insomnia, irritability, renal damages and weakness of muscles (Naseem and Tahir, 2001). Chromium exits in the aquatic environment mainly in two states: Cr^{3+} and Cr^{6+}. In general, Cr^{6+} is more toxic than Cr^{3+}. Cr^{6+} affects human physiology, accumulates in the food chain and causes severe health problems ranging from simple skin irritation to lung carcinoma (Khezami and Capart, 2005).

The pollution problem occurring in the Nyabugogo wetland raised many concerns in the people living around the wetland, the local authority and the Rwandan government in particular. As reported by *THE NEWTIMES*, a Rwandan daily newspaper article published on the 9th October 2009, the government has decided to relocate close to 100 commercial entities located in the Gikondo wetland to another industrial park in the Free Trade Zone in the Gasabo district. This measure was mainly motivated by the fact that the firms were found to be poorly constructed and hazardous to the surrounding environment. According to an environmental impact assessment carried out by the Rwanda Environment Management Authority (REMA), the entities were wrongly positioned and were contaminating wetlands. The Director General of REMA said that once the industries are moved, water scarcity in Kigali will be minimized. "If the barriers to wetlands were removed, there wouldn't be water scarcity. Moreover, the wetlands filter, store water and prevent floods". Furthermore, based on the law n° 04/2005 of 08 / 04 / 2005 determining the modalities of protection, conservation and promotion of the environment in Rwanda, It was clear that, that decision would be taken in order to stop further deterioration of that natural wetland and plan for a restoration of the destroyed area.

7.2. Sustainable technology for metal removal

Several treatment methods have been developed for removal of heavy metal from wastewater. Most of the treatment methods are identified as physico-chemical processes like adsorption (Jusoh et al., 2007; Kang et al., 2008; Apiratikul and Pavasant, 2008a; Al-Jlil and Alsewailem, 2009; Gu and Evans, 2008; Sekomo et al., 2012), membrane filtration (Landaburu-Aguirre et al., 2009; Li et al., 2009; Trivunac and Stevanovic, 2006; Kim et al., 2005), ion-exchange (Motsi et al., 2009; Taffarel and Rubio, 2009; Doula and Dimirkou, 2008), bioreactors (Lens et al., 2002; Van Hullebusch et al., 2006; Lenz et al., 2008; Kousi et al., 2007; Alvarez et al., 2007; Sekomo et al., submitted), electro coagulation and electro flotation (Lai and Lin, 2003, 2004; Can et al., 2006).

Traditionally, heavy metals removal from wastewater has been carried out by chemical precipitation due to the simplicity of the process and its inexpensive capital cost. However, chemical precipitation is usually well suited for the treatment of wastewaters with high heavy metal concentrations. It is ineffective when the metal ion concentration is low. Furthermore, chemical precipitation is not economical and produces in general large amounts of sludge, requesting additional treatment steps. Ion exchange has been widely applied for the removal of heavy metals from wastewater. However, ion-exchange resins must be regenerated by chemical reagents when they are exhausted and the regeneration can cause serious secondary pollution. And it is expensive, especially when treating a large amount of wastewater containing heavy metals in low concentration, so they cannot be used at large scale.

Adsorption is a recognized method for the removal of heavy metals from low concentration wastewater containing heavy metals. The high cost of activated carbon limits its use in adsorption. Many varieties of low-cost adsorbents have been developed and tested to remove heavy metal ions. However, the adsorption efficiency depends on the type of adsorbent. Biosorption of heavy metals from aqueous solutions is a relatively new process that has proven very promising for the removal of heavy metals from wastewater. The membrane filtration technology can remove heavy metal ions with high efficiency, but its problems such as high cost, process complexity, membrane fouling and low permeate flux have limited their use in heavy metal removal. Using coagulation flocculation for heavy metal removal from wastewater, the produced sludge has good sludge settling and dewatering characteristics. But this method involves chemical consumption and increased sludge volume generation.

Flotation offers several advantages over the more conventional methods, such as high metal selectivity, high removal efficiency, high overflow rates, low detention periods, low operating cost and production of more concentrated sludge (Rubio et al., 2002). But the disadvantages involve high initial capital costs, high maintenance and operation costs. Electrochemical heavy metal wastewater treatment techniques are regarded as rapid and well-controlled that require fewer chemicals, provide good reduction yields and produce less sludge. However, electrochemical technologies involve high initial capital investments and expensive electricity supply, this restricts its development. Although all above techniques can be employed for the treatment of heavy metal contaminated wastewater, it is important to mention that the selection of the most suitable treatment techniques depend on the initial metal concentration, the other components present in the wastewater, capital investment and operational cost, plant flexibility and reliability, environmental impact, etc. (Kurniawan et al., 2006).

Although all these techniques can be used in heavy metal removal, they have their inherent advantages and limitations. For developing countries, there is a problem in opting for high-tech methods because these require high energy input, skilled personnel for operation and maintenance. Furthermore, they also generate large amounts of chemical waste or will need a regeneration step involving chemicals. For these reasons, alternative solutions not requiring high investment costs, skilled people and local available materials are needed. Biological metal sulfide precipitation and phytoremediation are alternative methods that, when applied in combination, do not involve high capital investment. Biological metal sulfide precipitation can be applied with the use of locally available carrier materials (volcanic rock) in a subsurface flow constructed wetland or by use of bioreactors. The effluent flowing out of the system can be further treated by phytoremediation as polishing step for metal removal as the plant themselves can't treat wastewater containing high metal concentrations.

7.3. Characterization of volcanic rock

The Gisenyi volcanic rock showed affinity for heavy metals and was therefore used as adsorbent for heavy metal removal from aqueous solution (Sekomo et al., 2012). As reported for many other solid adsorbents (Alemayehu and lennartz, 2009; Buamah et al., 2008; Ahmet et al., 2007; Bosso and Enzweiler, 2002), the removal efficiency is affected by parameters such as the contact time, the pH, the adsorbent dosage, the heavy metal concentration and the competition with other heavy metals. The adsorption increases with pH and adsorbent dosage. It was found that the pseudo second-order kinetic model fitted very well the experimental data and therefore suggests a removal mechanism of chemical nature. Among many low cost adsorbents (Peric et al., 2004; Prasad et al., 2008; Alemayehu et al., 2009), the Gisenyi volcanic rock was found to be also a promising low cost and abundant adsorbent for heavy metal removal from polluted water. Furthermore, it has also the advantage of being a stable sorbent that can be reused in multiple sorption-desorption-regeneration cycles (Chapter 4). This has also been reported by other researchers (Hong et al., 1999; Beolchini et al., 2003; Sekhar et al., 2004; Hammaini et al., 2007; Mata et al., 2010; Katsou and Tzanoudaki, 2010) investigating other adsorbent materials.

Based on the monthly volume consumption of 7,000 m^3 of clean water, the textile industry consumes a daily volume of 350 m^3 = 350,000 L. Based on the laboratory scale setup with the following parameters: Adsorbent mass = 16.2 kg; Column length = 2 m; column diameter = 14 cm; HRT = 1 day; effective volume (V_e) = 22.4 L; flow rate (R_f) = 31.16 ml/min. and a capacity of treating 45 L/day; extrapolating these data, with the aim of treating 350,000 L/day, we will need to upgrade parameters of the laboratory scale set up by multiplying them with a factor of 350,000/45 = 7,777.8. Looking only at the adsorbent side, the system will require a quantity of 126,000 kg of volcanic rocks, for treating the wastewater. Even if laboratory scale studies showed effectiveness of the rock in heavy metal removal, the application of this process at industrial scale in general will not be possible due to (1) the large quantity of adsorbent material that will be required for the process and (2) the saturation of the adsorbent that will involve subsequent running costs and extra time that would be required for the regeneration of the adsorbent prior further reuse of the material

(Sekomo et al. 2012a). In order to address these issues, the introduction of metal sulfide precipitation as alternative long term and sustainable mechanism for heavy metal removal is very important to avoid the use of huge quantities of adsorbent and the regeneration of the adsorbent.

7.4. Sulfide precipitation in Upflow Anaerobic Packed Bed reactors

The study of the removal of heavy metals by natural systems by means of natural wetlands showed that sediment was the most important sink for heavy metals (Chapter 2). Furthermore, the adsorption of heavy metal by volcanic rock as adsorbent (Chapter 4) showed that the adsorption worked however; the saturation of the adsorbent was a major issue. Therefore, the introduction of another technology for heavy metal removal was needed. The removal of heavy metals by sulfide precipitation showed an important increase in the removal efficiency (Chapter 5). This study showed that heavy metals have been efficiently removed (> 90 %) in Upflow Anaerobic Packed Bed reactors. The mass balance analysis (Chapter 4) showed that metals were mainly removed by accumulation in the sludge (61.9 % – 84.7 % of total removal), probably as metal sulphide precipitates. Furthermore, the use of porous volcanic rock adsorbents played an additional role in the metal removal by adsorption combined with biosorption, contributing with 6.2 % – 32.3 % to the overall removal.

From the mass balance analysis equation (Chapter 4), heavy metals have been removed from the system by sulphide precipitation and adsorption combined with biosorption onto the adsorbent. It would be interesting if the metal speciation analysis of the effluent from the column system was conducted. This would provide additional information on the metal species present in the effluent and allowed further comprehension and interpretation regarding the fate of these toxic compounds. It is widely recognized that the distribution, mobility and bioavailability of heavy metals in the environment depend not only on their total concentration but also on the association form in the solid phase to which they are bound (Filgueiras et al., 2002; Lasheen and Ammar, 2009). As reported by Stoveland et al., 1979, the removal of heavy metal is altered by metal complexation. Furthermore, metals can be released from complexes or desorbed from particules, especially if the pH of the receiving water is different than pH of the effluent (Shi et al., 1998) or if changes in redox potential also occur (). In this study, no chemical speciation has been used. However, the performance of the reactor was assessed by looking at the COD and SO_4^{2-} removal. Furthermore, the monitoring of COD, EC, pH, and redox potential parameters in the pore water elucidates the removal mechanisms. Results have shown sulphide precipitation as the major and long term removal mechanism for heavy metals. However, heavy metals were present in the effluent. Therefore, this requires an additional removal step in order to reach the effluent discharge limit for heavy metals in the environment.

Besides the toxicity of heavy metal, organic compounds present in textile effluent are also toxic to the SRB. Many dyes are also made from known carcinogens, such as benzidine and other aromatic compounds, all of which might be reformed as a result of microbial metabolism (Clarke and Anliker, 1980). With particular reference to the textile industry, dyes are classified as anthraquinone, phthalocyanine, triphenylmethane, heterocyclic, and azo, with the last mentioned representing one of the largest class used and characterized by the functional azo group (-N=N-). These brightly colored dyes are water-soluble and are extremely resistant to microbial and physicochemical degradation, including conventional processes of wastewater treatment. The microbiological decolorization of industrial effluent wastewater containing these dyes is ongoing with an increasing number of studies being reported (Man & Chaudhari, 2002; Mendez-Paz et al., 2004; Ambrosio & Campos-Takaki, 2004; Kumar et al., 2005; Mutambanengwe et al., 2007). The loss in color, for a typical azo dye, is due to a reduction of the azo group first to the bis-amine then to two separate amines. These dyes and their degradative aromatic amine products are toxic and mutagenic to living systems and the environment in general.

Applying this technique at industrial scale, an upgrade of design parameters is needed as conducted in point 7.4. However, in this case not as much adsorbent mass will be required as for the adsorption process because the sulfide precipitation is the major removal mechanism in the reactor. Taking into account the volume of 350,000 L consumed per day, the factory requires a bioreactor that can approximately handle a volume of 14, 600 L / hour. Based on the total volume of the laboratory scale reactor 30.772 L (V_t) and the effective volume 22.4 L (V_e), the volume occupied by the adsorbent in the column was: $V_t - V_e = V_{v.r} \Rightarrow V_{v.r} = 30.772 - 22.4 = 8.373\ L$. Extrapolating these data to a real scale and assuming that a bioreactor of a total volume of 15,000 L is required, the volume occupied by the adsorbent will be equal to:

$$V_{v.r} = \frac{8.372 \times 15,000}{30.772} = 4,080.98\ L = 4.08 m^3$$

This volume will correspond approximately to:

$$Mass\ adsorbent = \frac{16.2\ kg \times 4.08\ m^3}{8.37 \times 10^{-3} m^3} = 7,895\ kg$$

7.5. Integrated system for wastewater treatment

Developing countries need ecological and cost effective technologies for wastewater treatment. In this study, an integrated system composed of a combination of two complementary mechanisms, the metal sulfide precipitation in a bioreactors and the phytoremediation by mean of a water hyacinth pond has been investigated. This system aimed at addressing the major shortcoming that macrophyte plants are suffering in wastewater treatment systems due to "metal accumulation and uptake by plants play a minor role in wetlands and ponds systems (Mays and Edwards, 2001)".

Metal sulfide precipitation is known as a long term heavy metal removal mechanism in natural systems for wastewater treatment as long as reducing condition are maintained (Lesage et al., 2007a; Sekomo et al., 2011b). Therefore, the application of this mechanism as a pretreatment in a wastewater treatment system will remove > 90 % of the pollutant charge from the wastewater; then the application of macrophyte plants can proceed as polishing step for the treatment system. Results from this study have shown that the bioreactor was mainly accumulating heavy metals in the inlet columns when compared to the outlet columns (Chapter 5). For a long term running, it is clear that the accumulation of metal rich sludge will be another issue that needs to be addressed. Removal of a pollutant from one phase to another is here beneficial as long as the latter phase is handled appropriately in order to avoid pollution spreading in the environment. The immobilization of metals in the sludge requires a further treatment that will ensure the immobilization even after the de-sludging process. This metal polluted sludge can be used for brick or concrete making for a further immobilization.

Metal concentrations in the water hyacinth plants showed high concentration on the roots system when compared to the stem and leaves. This finding agrees with many studies in natural and constructed wetlands showing that adsorption on the roots system was the predominant mechanism compared to uptake and translocation (Judith and Pedrik, 2004; Marchand et al., 2010; Sekomo et al., 2011a). Heavy metal concentrations in the water hyacinth increased with increasing pollutant concentration in the wastewater. However, the concentrations in the stem and leaves were less and not correlated with concentrations found at the root system of the plant. In general, low percentages of the mass loadings of heavy metals were accumulated in the water hyacinth when looking at the whole system. Thus, confirming the use of plants as polishing step in the natural systems for wastewater treatment.

7.6. Duckweed and algae ponds as alternatives for the water hyacinth

Macrophyte plants have shown their capacity in heavy metal removal from wastewater (Rai et al., 1995; Denny et al., 1995; Mungur et al., 1997; Zhihong et al., 1998; Obarska, 2001; Cheng et al., 2002; Keskinkan, 2005). Their application is limited by the pollutant charge as macrophyte plants have been reported to play a lesser role in heavy metal removal in a wetland (Vymazal, 1995; Murray-Gulde et al., 2005; Vymazal and Krasa, 2005; Kadlec and Wallace, 2008; Sekomo et al., 2011a). In this study, three macrophyte plants (Algae, Duckweed and Water hyacinth) have been selected due to their fast growth (Duckweed and Water hyacinth), the presence of an elaborate roots system (Water hyacinth) and the change in dissolved oxygen it induces within the system (Algae). These differences between the three species could lead to differences in treatment performance. In this study, algae and duckweed based ponds (Rousseau et al. 2011; Sekomo et al. 2012b) were investigated in order to compare their treatment performance as alternative for water hyacinth. Water hyacinths are very productive photosynthetic plants. Their rapid growth has been reported as a serious nuisance problem in many slow flowing southern waterways (Reed et al., 1997). However, these same attributes become an advantage when used in a wastewater treatment system; where its elaborate roots system provide an attachment surface for bacterial growth, therefore enabling good contact between wastewater and the macrophyte plant.

Natural systems specifically macrophyte plants, have been used as an alternative cost-effective technology. It has been shown that they can efficiently remove heavy metals from wastewater (Rodgers and Dunn, 1992; Lakatos et al., 1997; Le Duc and Terry, 2005). Plants contribute in these systems for the removal of pollutants, especially nutrients, BOD and heavy metal removal. They also host micro-organisms which in turn provide sites for metal sorption as well as carbon sources for bacterial metabolism (Jacob and Otte, 2004; Marchand et al., 2010). Water hyacinth (*Eichhornia crassipes*) and duckweed (*Lemna sp.*) are commonly used in aquatic treatment systems. These plants and algae influence the redox and pH conditions of the aquatic systems as a result of photosynthesis and respiration processes (Shilton, 2005). Therefore, they also contribute to the metal removal processes. Given that pH and redox fluctuations are much higher in algal ponds than in duckweed ponds, one may expect different metal removal efficiencies in these pond systems. However, results from this study did not show the difference expected in metal removal based on the pond type. In general, the overall performance was close for both ponds (Chapter 6), showing that duckweed and algal ponds are both suited as polishing step for heavy metal removal at lower concentration.

In summaryzing, the table below compares removal efficiencies of each individual pond system, algae, duckweed and water hyacinth. In general, all these ponds system show one advantage of removing heavy metal in polluted water. Algae and duckweed ponds showed relatively close removal efficiencies in that regards. The water hyacinth pond showed the highest removal efficiencies when compared to algae and duckweed ponds. It can therefore be consider as a suitable pond system to be used as polishing step for wastewater containing heavy metal. However, it major disadvantage is that this plant is known to be invasive and is causing problems in the environment if not controlled. The disadvantage of algae and duckweed ponds is the large space that will be required to setup a system of many ponds in series in order to reach the desired performance in metal removal.

	Algae pond	Duckweed pond	Water hyacinth pond	UAPB reactor	Integrated system
Cd	22 %	24 %	61 %	97 %	98 %
Cr	98 %	95 %	-	-	-
Cu	25 %	22 %	59 %	99 %	99%
Pb	37 %	36 %	49 %	97 %	98%
Zn	62 %	59 %	42 %	91 %	84 %

The UAPBR system has the major advantage of removing heavy metal at very satisfactory level > 90 %. However, it has one major disadvantage the release of hydrogen sulphide gas creating nuisance in the sourrounding of the system. This requires additional measure to prevent the release of the bad smell in the environement. The combination of each pond system with the UAPBR has the advantage of increasing further the removal of heavy metal. However, it also presents the disadvantage due to the toxicity effect of the excess of hydrogen sulphide to algae and plants in the pond system requiring also additional measure of decreasing sulphate concentration in the reactor to keep the integrated syetm fully operational.

7.7. Conclusions

This PhD study investigated metal removal mechanisms in natural wastewater treatment systems. The most important conclusions are:

➢ This study contributed to the understanding of mechanisms for heavy metal removal using anaerobic bioreactors and natural systems and used this to develop an integrated system. Moreover, it also presents how natural systems can be complementary in the removal of heavy metals as a post treatment system.
➢ The field survey in a natural wetland receiving textile wastewater demonstrated that heavy metals are mainly removed by sedimentation. However, adsorption onto the roots system, uptake and translocation in the macrophyte plants also contributed to the heavy metal removal.

- Laboratory scale adsorption tests of heavy metal (Cd, Cu, Pb and Zn) to a locally available adsorbent (Gisenyi volcanic rock) have been conducted. Maximum adsorption capacity and kinetic studies have been reported. Furthermore, the adsorbent can be regenerated for further reuse.
- Metal sulfide precipitation has been tested at laboratory scale as long term removal mechanism for heavy metals. Removal efficiencies exceeding 90 % were reached by the bioreactor.
- The integrated system for heavy metal removal showed how two complementary systems for heavy metal removal can work in combination and thus achieved a good removal performance. This system is highly recommended to the textile industry and other industries facing the problem of heavy metal pollution as low cost treatment option for their wastewater.
- The use of algal and duckweed ponds as alternative for water hyacinth plants was showing no differences in heavy metal removal based on abiotic differences. Both systems performed equally and are well suited as polishing step for wastewater containing low metal concentration.

7.8. Future perspectives

- It is recommended that the monitoring of heavy metals in the Nyabugogo wetland is further conducted to investigate the human health impact of using wetland water for crop irrigation and fish farming.
- It is also recommended to investigate the use of the volcanic rock as attachment surface for bacterial growth in domestic wastewater treatment for BOD removal.
- It is recommended that long term experiments with the integrated system combining the bioreactor packed with the volcanic rock are conducted to assess its performance limit. Also experiments with the UAPB reactors with algae, duckweed and water hyacinth ponds mounted in parallel are required for an in depth comparison between those ponds treatment systems. This will give an evaluation about the toxicity effect of sulfide from the bioreactor to the algae, duckweed or water hyacinth in the pond systems.
- It is aso recommended to conduct experiments with real textile wastewater in order to evaluate the efficiency of the system at industrial site. Furthermore, the effect of dyes will then also be evaluated at the overall performance of the system.
- It is recommended that further investigation of the potential use of the Gisenyi volcanic rock for Fe and Mn removal is carried out for future use as adsorbent material for drinking water treatment.

7.9. References

Ahmet, S., Mustafa, T. & Mustafa, S. (2007). Adsorption of Pb(II) and Cr(III) from aqueous solution on Celtek clay. Journal of Hazardous Materials 144, 41 – 46.

Alemayehu, E. & Lennartz, B. (2009). Virgin volcanic rocks: Kinetics and equilibrium studies for the adsorption of cadmium from water. Journal of Hazardous Materials 169, 395 – 401.

Al-Jlil, S.A. & Alsewailem, F.D. (2009). Saudi Arabian clays for lead removal in wastewater. Applied Clay Science 42, 671 – 674.

Alvarez, M. T., Crespo, C. & Mattiasson, B. (2007). Precipitation of Zn(II), Cu(II) and Pb(II) at bench-scale using biogenic hydrogen sulfide from the utilization of volatile fatty acids. Chemosphere 66, 1677 – 1683.

Ambrosio, S.T. & Campos-Takaki, G.M. (2004). Decolourisation of reactive azo dyes by *Cunninghamella elegans* UCP 542 under co-metabolic conditions. Bioresource Technology 91, 69 – 75.

Apiratikul, R. & Pavasant, P. (2008a). Sorption of Cu^{2+}, Cd^{2+}, and Pb^{2+} using modified zeolite from coal fly ash. Chemical Engineering Journal 144, 245 – 258.

Beolchini, F., Pagnanelli, F., Toro, L. & Veglio, F. (2003). Biosorption of copper by Sphaerotilus natans immobilised in polysulfone matrix: equilibrium and kinetic analysis, Hydrometallurgy 70, 101 – 112.

Bosso, S.T. & Enzweiler, J. (2002). Evaluation of heavy metal removal from aqueous solution onto scolecite. Water Research 36 (19), 4795 – 4800.

Buamah, R., Petruseveski, B. & Schippers, J.C. (2008). Adsorptive removal of manganese(II) from the aqueous phase using iron oxide coated sand. Journal *of* Water Supply. Research and Technology 57, 1 – 11.

Can, O. T., Kobya, M., Demirbas, E. & Bayramoglu, M. (2006). Treatment of the textile wastewater by combined electrocoagulation. Chemosphere 62, 181 – 187.

Cheng, J., Bergamann, B.A., Classen, J.J., Stomp, A.M. & Howard, J.W. (2002). Nutrient recovery from swine lagoon water by spirodela punctata. Bioresource Technology 81, 81–85

Clarke, E.A. & Anliker, R. (1980). Organic dyes and pigments. In The Handbook of Environmental Chemistry, Vol. 3, Part A. Anthropogenic Compounds, ed. O. Hutzinger. Springer, Heidelberg, pp. 181 – 215.

Denny, P. (1995). Heavy metal contamination of Lake George (Uganda) and its wetlands. Hydrobiologia, 257, 229 – 239.

Denny, P., Bailey, R., Tukahirwa, E. & Mafabi, P. (1995). Heavy metal contamination of Lake George (Uganda) and its wetlands. Hydrobiologia, 297(3), 229 – 239.

Doula, M.K. & Dimirkou, A. (2008). Use of an iron-overexchanged clinoptilolite for the removal of Cu^{2+} ions from heavily contaminated drinking water samples. Journal of Hazardous Materials 151, 738 – 745.

EC (2001). Commission Regulation as regards heavy metals, Directive 2001/22/EC, No: 466/2001.

Fenglian, F. & Wang, Q. (2011). Removal of heavy metal ions from wastewaters: A review. Journal of Environmental Management 92. 407 – 418.

Filgueiras, A.V., Lavilla, I. & Bendicho, C. (2002). Chemical sequential extraction for metal partitioning in environmental solid samples, Journal of Environmental Monitoring 4, 823 – 857.

Gu, X.Y. & Evans, L.J. (2008). Surface complexation modelling of Cd(II), Cu(II), Ni(II), Pb(II) and Zn(II) adsorption onto kaolinite. Geochimica et Cosmochimica Acta 72, 267 – 276.

Hammaini, A., Gonzalez, F., Ballester, A., Blazquez, M.L. & Munoz, J.A. (2007). Biosorption of heavy metals by activated sludge and their desorption characteristics. Journal of Environmental Management 84, 419 – 426.

Hong, P.K.A., Li, C., Banerjiand, S.K. & Regmi, T. (1999). Extraction, recovery and biostability of EDTA for remediation of heavy metal contaminated soil. Journal of Soil Contamination 8, 81 – 103.

Jacob, D.L. & Otte, M.L. (2004). Influence of Typha latifolia and fertilization on metal mobility in two different Pb–Zn mine tailings types. Science of the Total Environment 333, 9 – 24

Judith, S.W. & Peddrick, W. (2004). Metal uptake, transport and release by wetland plants: Implications for phytoremediation and restoration. Environment International, 30, 685 – 700.

Jusoh, A., Shiung, L.S., Ali, N. & Noor, M.J.M.M. (2007). A simulation study of the removal efficiency of granular activated carbon on cadmium and lead. Desalination 206, 9 – 16.

Jones, M. & Humphries, S. (2002). Impacts of the C4 Sedge Cyperus papyrus L. on carbon and water fluxes in an African Wetland. Hydrobiologia 488, 107 – 113.

Kadlec, R.H. & Wallace, S.D. (2008). Treatment Wetlands second edition, CRC Press.

Kang, K.C., Kim, S.S., Choi, J.W. & Kwon, S.H. (2008). Sorption of Cu^{2+} and Cd^{2+} onto acid- and base-pretreated granular activated carbon and activated carbon fiber samples. Journal of Industrial and Engineering Chemistry 14, 131 – 135.

Kansiime, F., Saunders, M.J. & Loiselle, S.A. (2007). Functioning and dynamics of wetland vegetation of Lake Victoria: an overview. Wetlands Ecology and Management 15, 443 – 451.

Katsou, E. & Tzanoudaki, M. (2010). Regeneration of natural Zeolite polluted by lead and zinc in wastewater treatment systems, Journal of Hazardous Materials, doi:10.1016/j.jhazmat.2010.12.061

Khezami, L. & Capart, R. (2005). Removal of chromium(VI) from aqueous solution by activated carbons: kinetic and equilibrium studies. Journal of Hazardous Materials 123, 223 – 231.

Kim, H.J., Baek, K., Kim, B.K. & Yang, J.W. (2005). Humic substance-enhanced ultrafiltration for removal of cobalt. Journal Hazardous Materials 122, 31 – 36.

Kousi, P., Remoudaki, E., Hatzikioseyian, A. & Tsezos, M. (2007). A study of the operating parameters of a sulphate-reducing fixed-bed reactor for the treatment of metal-bearing wastewater. In: 17th International Biohydrometallurgy Symposium, Germany, Frankfurt am Main.

Kumar, K., Devi, S.S., Krishnamurthi, K., Gampawar, S., Mishra, N., Pandya, G.H. & Chakrabarti, T. (2005). Decolourisation, biodegradation and detoxification of benzidene based azo dye. Bioresource Technology 97, 407 – 413.

Kurniawan, T.A., Chan, G.Y.S., Lo, W.H. & Babel, S. (2006). Physico-chemical treatment techniques for wastewater laden with heavy metals. Chemical Engineering Journal 118, 83 – 98.

Lai, C.L. & Lin, S.H. (2003). Electrocoagulation of chemical mechanical polishing (CMP) wastewater from semiconductor fabrication. Chemical Engineering Journal 95, 205 – 211.

Lai, C.L. & Lin, S.H. (2004). Treatment of chemical mechanical polishing wastewater by electrocoagulation: system performances and sludge settling characteristics. Chemosphere 54, 235 – 242.

Lakatos, G., Kiss, M. K., Kiss, M. & Juhasz, P. (1997). Application of constructed wetlands for wastewater treatment in Hungary. Water Science and Technology 35(5), 331 – 336.

Lasheen, M.R. & Ammar, N.S. (2009). Assessment of metals speciation in sewage sludge and stabilized sludge from different Wastewater Treatment Plants, Greater Cairo, Egypt. Journal of hazardous materials 164, 740 – 749.

Le Duc, D.L. & Terry, N. (2005). Phytoremediation of toxic trace elements in soil and water. Journal of Industrial Microbiology and Biotechnology 32, 514 – 520.

Lens, P.N.L., Vallero, M., Esposito, G. & Zandvoort, M. (2002) Perspectives of sulphate reducing bioreactors in environmental biotechnology. Reviews in Environmental Science and Biotechnology 1 (4), 311 – 325.

Lenz, M., Van Hullebusch, E.D., Hommes, G., Corvini, P.F.X. & Lens, P.N.L. (2008) Selenate removal in methanogenic and sulphate-reducing upflow anaerobic sludge bed reactors. Water Research 42, 2184 – 2194.

Lesage, E., Rousseau, D.P.L., Meers, E. Tack, F.M.G. & De Pauw, N. (2007a). Accumulation of metals in a horizontal subsurface flow constructed wetland treating domestic wastewater in Flanders, Belgium. Science of the Total Environment 380, 102 – 115.

Man, B. & Chaudhari, S. (2002). Anaerobic decolourisation of simulated textile waste water containing azo dyes. Bioresource Technology 82, 225 – 231.

Manios, T., Stentiford, E.I. & Millner, P.A. (2003). The effect of heavy metals accumulation on the chlorophyll concentration of Typha latifolia plants, growing in a substrate containing sewage compost and watered with metaliferous water. Ecological Engineering, 20, 65 – 74.

Marchand, L., Mench, M., Jacob, D.L. & Otte, M.L. (2010). Metal and metalloid removal in constructed wetlands, with emphasison the importance of plants and standardized measurements: A review. Environmental Pollution 158, 3447 – 3461

Mata, Y.N., Blazquez, M.L., Ballester, A., Gonzalez, F. & Munoz. J.A. (2010). Studies on sorption, desorption, regeneration and reuse of sugar-beet pectin gels for heavy metal removal. Journal of Hazardous Materials 178, 243 – 248.

Mays P.A. & Edwards G.S. (2001). Comparison of heavy metal accumulation in a natural wetland and constructed wetlands receiving acid mine drainage, Ecological Engineering 16, 487 – 500.

Mendez-Paz, D., Omil, F. & Lema, J.M. (2004). Anaerobic treatment of azo dye Acid Orange 7 under batch conditions. Enzyme and Microbial Technology 36, 264 – 272.

Monday, S.L., Kansiime, F., Denny, P. & James, S. (2003). Heavy metals in Lake George, Uganda, with relation to metal concentrations in tissues of common fish species. Hydrobiologia, 499, 83 – 93.

Motsi, T., Rowson, N.A. & Simmons, M.J.H. (2009). Adsorption of heavy metals from acid mine drainage by natural zeolite. International Journal of Mineral Processing 92, 42 – 48.

Mungur, A.S., Shutes, R.B.E., Revit, D.M. & House, M.A. (1997). An assessment of metal removal by a laboratory scale wetland. Water Science and Technology 35, 125 – 133.

Murray-Gulde, C., Huddleston, G.M. Garber, K.V. & Rodgers J.H. (2005). Contributions of Schoenoplectus californicus in a constructed wetland system receiving copper contaminated water. Water Air and Soil Pollution 163(1-4), 355 – 378.

Mutambanengwe, C.C.Z., Togo, C.A. & Whiteley, C.G. (2007). Decolorization and Degradation of Textile Dyes with Biosulfidogenic Hydrogenases. Biotechnology Progress 23, 1095 – 1100.

Naseem, R. & Tahir, S.S. (2001). Removal of Pb(II) from aqueous solution by using bentonite as an adsorbent. Water Research. 35, 3982 – 3986.

Obarska, H. & Pempkowiak, (2001). Retention of selected heavy metals: Cd, Cu, Pb in a hybrid wetland system. Water Science and Technology 44(11–12), 463 – 468

Oyaro, N., Juddy, O., Murago, E.N.M. & Gitonga, E. (2007). The contents of Pb, Cu, Zn and Cd in meat in Nairobi, Kenya. Int. J. Food Agric. Environ. 5, 119 – 121.

Paulino, A. T., Minasse, F. A. S., Guilherme, M. R., Reis, A. V., Muniz, E. C., & Nozaki, J. (2006). Novel adsorbent based on silkworm chrysalides for removal of heavy metals from wastewaters. Journal of Colloid and Interface Science 301, 479 – 487.

Peric, J., Trgo, M. & Medvidovic, N.V. (2004). Removal of zinc, copper, and lead by natural Zeolite: a comparison of adsorption isotherms. Water Research 38 (7), 1893 – 1899.

Prasad, M., Xu, H.Y. & Saxena, S. (2008). Multi-component sorption of Pb(II), Cu(II) and Zn(II) onto low-cost mineral adsorbent. Journal of Hazardous Materials 154, 221 – 229.

Rai, U.N., Sarita, S., Tripathi, R.D. & Chandra, P. (1995). Wastewater treatability of some aquatic macrophytes: removal of heavy metals. Ecological Engineering 5, 5 – 12.

Reed, S.C., Middlebrooks E.J. & Crites R.W. (1997). Natural Systems for Waste Management and Treatment. McGraw-Hill Book Co. NY.

Rodgers, J.H.Jr. & Dunn, A. (1992). Developing design guidelines for constructed wetlands to remove pesticides from agricultural runoff. Ecological Engineering 1, 83 – 95.

Rousseau, D.P.L., Sekomo, C.B., Saleh, S.A.A.E. & Lens, P.N.L. (2011). Duckweed and Algae Ponds as a Post-Treatment for Metal Removal from Textile Wastewater. Water and nutrient management in natural and constructed wetlands. 63 – 75,

Rubio, J., Souza, M. L., & Smith, R. W., (2002). Overview of flotation as a wastewater treatment technique. Mineral Engineering 15, 139 – 155.

Sekhar, K.C., Kamala, C.T., Chary, N.S., Sastry, A.R.K., Rao, T.N. & Vairamani, M. (2004). Removal of lead from aqueous solutions using an immobilized biomaterial derived from a plant biomass, Journal of Hazardous Materials 108, 111 – 117.

Sekomo, C.B., Nkuranga, E., Rousseau, D.P.L. & Lens, P.N.L. (2011a). Fate of Heavy Metals in an Urban Natural Wetland: The Nyabugogo Swamp (Rwanda). Water Air and Soil Pollution 214, 321 – 333.

Sekomo, C.B., Rousseau, D.P.L. & Lens, P.N.L. (2012). Use of Gisenyi Volcanic Rock for Adsorptive Removal of Cd(II), Cu(II), Pb(II), and Zn(II) from Wastewater. Water Air and Soil Pollution 223; 533 – 547.

Sekomo, C.B., Kagisha, V., Rousseau, D.P.L. & Lens, P.N.L. (2011b). Heavy Metal Removal by Combining Anaerobic Upflow Packed Bed Reactors with Water Hyacinth Ponds. Environmental technology. DOI:10.1080/09593330.2011.633564.

Sekomo, C.B., Rousseau, D.P.L., Saleh, S.A.A.E. & Lens, P.N.L. (2012). Heavy metal removal in duckweed and algae ponds as a polishing step for textile wastewater treatment. Submitted to Ecological Engineering.

Shi, B., Allen, H.E. & Grassi, M.T. (1998). Changes in dissolved and particulate copper following mixing of POTW effluents with Delaware river water. Water Research 32, 2413–2421.

Shilton, A. (2005). Pond Treatment Technology. London, UK, IWA Publishing. ISBN: 1843390205.

Stoveland, S., Perry, R. & Lester, J.N. (1979). The influence of nitrilotriacetic acid on heavy metal transfer in the activated sludge process-II at varying and shock loadings. Water Research 13, 1043 – 1054.

Taffarel, S.R. & Rubio, J. (2009). On the removal of Mn^{2+} ions by adsorption onto natural and activated Chilean zeolites. Mineral Engineering 22, 336 – 343.

The New Times – Rwandas First Daily. 100 firms to be evicted from Gikondo wetland. http://www.newtimes.co.rw/index.php?issue=14043 (Accessed on the 9/09/2009)

Trivunac, K. & Stevanovic, S. (2006). Removal of heavy metal ions from water by complexation-assisted ultrafiltration. Chemosphere 64, 486 – 491.

Van Hullebusch, E.D., Gieteling, J., Zhang, M., Zandvoort, M.H., Daele, W.V., Defrancq, J. & Lens, P.N.L. (2006). Cobalt sorption onto anaerobic granular sludge: isotherm and spatial localization analysis. Journal of Biotechnology. 121 (2), 227 – 240.

Vymazal, J. & Krasa P. (2005). Heavy metals budget for constructed wetland treatment municipal sewage. In: Natural and Constructed Wetlands – Nutrients, Metals and Management, Vymazal J. (ed) Backhuys Publishers: Leiden, The Netherlands, 135 – 142.

Vymazal, J. (1995). Algae and nutrient cycling in wetlands. CRC Press / Lewis Publishers: Boca Raton, Florida.

Wang, J., Huang, C.P. & Allen, H.E. (2006). Predicting metals partitioning in wastewater treatment plant influents. Water Research 40, 1333 – 1340.

Williams, L.E., Pittman, J.K. & Hall, J.L. (2000). Emerging mechanisms for heavy metal transport in plants. Biochimica et Biophysica Acta, 1465 (12), 104 – 126.

Zhihong, Y., Alan, J.M.B., Ming-Hung, W. & Arthur, J.W. (1998). Zinc, lead and cadmium accumulation and tolerance in Typha latifolia as affected by iron plaque on the root. Aquatic botany 61, 55 – 67.

Samenvatting

Economische ontwikkeling, verstedelijking en bevolkingsgroei zijn drie parameters die momenteel de waterkwaliteit beïnvloeden. Textiel afvalwater is een typisch industrieel afvalwater dat ontstaat door menselijke activiteit. Dit alkalisch afvalwater bevat vele polluenten en heeft een hoge BZV en CZV belasting. Polluenten in textiel afvalwater zijn onder andere zwevende stoffen, minerale olie (bijv. antischuim producten, vet, spinnerij smeermiddelen), niet-biodegradeerbare of laag-biodegradeerbare oppervlakte-actieve stoffen en andere organische stoffen waaronder fenolen afkomstig van natte afwerkingsprocessen (bijv. verven) en gehalogeneerde koolwaterstoffen afkomstig van solventen gebruikt bij het bleken. Effluent van verfprocessen heeft een kleur en bevat vaak significante concentraties aan zware metalen (bijv. chroom, koper, zink, lood of nikkel). Dit onderzoek focust zich op vervuiling veroorzaakt door zware metalen in textielafvalwater omwille van hun toxiciteit, en omdat de meeste andere studies toegespitst waren op het verwijderen van organisch materiaal.

In Rwanda draagt textielafvalwater in belangrijke mate bij aan de vervuiling met zware metalen van de moerassen waarin het afvalwater geloosd wordt. Hoofdstuk 3 geeft een beoordeling van de metaal contaminatie in het Nyabugogo moeras waarin het textiel afvalwater van UTEXRWA (Kigali, Rwanda) geloosd wordt. In dit moeras worden allerhande afvalwaters van de stad zonder enige vorm van zuivering geloosd. Metaalvervuiling (Cd, Cr, Cu, Pb en Zn) werd onderzocht in alle milieucompartimenten van het moeras: water en sediment, de meest prominente plantensoorten in alle milieucompartimenten van het moeras: water en sediment, de meest prominente plantensoort *Cyperus papyrus*, vissen (*Clarias sp.* en *Oreochromis sp.*) en borstelwormen. Cr, Cu en Zn concentraties in het moeraswater waren door de band genomen lager dan de limiet opgenomen in de drinkwaterstandaard van de Wereld Gezondheidsorganisatie (WGO 2008), terwijl Cu en Pb consistent boven de limiet uitkwamen. Evenwel lagen alle metaalconcentraties, op die van Cd na, onder het maximum aanvaardbare niveau voor irrigatie. De hoogste metaalaccumulatie werd waargenomen in het sediment, met concentraties tot 4,2 mg/kg voor Cd, 68,0 mg/kg voor Cu, 58,3 mg/kg voor Pb en 188.0 mg/kg voor Zn, gevolgd door accumulatie in de wortels van *Cyperus papyrus* met concentraties tot 4,2 mg/kg voor Cd, 45,8 mg/kg voor Cr, 29,7 mg/kg voor Cu en 56,1 mg/kg voor Pb. Met uitzondering van Cu en Zn, werden hoge metaalconcentraties (Cd, Cr en Pb) aangetroffen in *Clarias sp.*, *Oreochromis sp.*, en in de borstelwormen. De gezondheid van mensen die water en producten van het moeras gebruiken, kan hierdoor in gevaar zijn.

De resultaten van hoofdstuk 3 tonen aan dat er een noodzaak is voor de ontwikkeling van goedkope, betrouwbare technologie die toegepast kan worden in ontwikkelingslanden voor het verwijderen van zware metalen uit afvalwater. Dit onderzoek streefde ernaar om de metaalverwijdering uit textielafvalwater te optimaliseren door het gebruik van een geïntegreerd systeem voor afvalwaterzuivering. Dit geïntegreerde systeem bestaat uit een combinatie van een anaeerobe reactor als hoofdzuivering, gevolgd door een vijver met macrofyten of algen als nazuiveringsstap. De hoofdzuivering verwijdert in principe de grootste hoeveelheid metalen door een combinatie van metaaladsorptie en metaalsulfide precipitatie. Nazuivering wordt bekomen door een fytoremediatie proces met algen of planten (eendenkroos en waterhyacint). Het geïntegreerde systeem zal de complementariteit van beide stappen aantonen voor wat betreft de verwijdering van zware metalen.

Op basis van de waargenomen metaalaccumulatie in de planten van het Nyabugogo moeras, werd beslist om ook een andere plantensoort te bestuderen, met name eendenkroos, en deze te vergelijken met algen, en dit omdat de abiotische condities sterk kunnen verschillen (Hoofdstuk 4). Twee zuiveringssystemen werden getest (algenvijvers en eendenkroosvijvers) op basis van de hypothese dat verschillende fysicochemische condities (pH, redoxpotentiaal en opgeloste zuurstof) kunnen leiden tot verschillen in metaalverwijdering. Deze systemen werden getest met een hydraulische verblijftijd van 7 dagen, onder twee verschillende metaalvrachten en lichtregimes (16/8 uur licht/donker versus 24 u licht). Resultaten tonen een Cr-verwijdering van 94% voor de eendenkroosvijvers en 98% voor de algenvijvers, onafhankelijk van de metaalvracht en het lichtregime. Onder het 16/8 lichtregime was er een goede verwijdering van Zn (~70%) bij een lage metaalbelasting, maar dit nam af tot minder dan 40% bij een hoge metaalbelasting. Bij diezelfde hoge metaalbelasting maar met een lichtregime van 24 u licht, nam de Zn verwijdering weer toe tot 80%. Pb, Cd en Cu vertoonden allen een gelijkaardig patroon, met verwijderingsefficiënties van 36% en 33% voor Pb, 33% en 21% voor Cd en 27% en 29% voor Cu voor de eendenkroosvijvers en algenvijvers, respectievelijk. Dit toont aan dat beide zuiveringssystemen niet bijzonder geschikt zijn als nazuiveringstechniek voor de verwijdering van deze metalen. Ondanks de significante verschillen in fysicochemische condities waren de verschillen in metaalverwijderingsefficiëntie tussen beide systemen eerder klein.

Gezien de tegenvallende resultaten van hoofdstuk 4, werden vervolgens alternatieve materialen voor metaalverwijdering gezocht. In dit kader werd een vulkanisch gesteente van Gisenyi in Noord–Rwanda gescreend als een potentieel adsorbens van metallische ionen (Cd, Cu, Pb en Zn) uit afvalwater. Adsorptie werd getest onder verschillende experimentele condities (initiële metaalconcentraties variërend van 1-50 mg/L, adsorbens dosis 4 g/L, vaste massa / vloeistof verhouding van 1:250, contact tijd 120 uur, deeltjesgrootte 250-900 μm). Adsorptie was optimaal bij een pH van 6. De maximum adsorptiecapaciteit was 6,23 mg/g, 10,87 mg/g, 9,52 mg/g en 4,46 mg/g voor Cd, Cu, Pb en Zn, respectievelijk. Het adsorptieproces vertoonde een snelle adsorptie gedurende de eerste 6 uur, gevolgd door een tragere adsorptie en tenslotte leidend tot een evenwicht na 24 uur. Er werd ook aangetoond dat deze vulkanische rots een stabiel adsorbens is die gedurende meerdere adsorptie-desorptie cycli kan worden gebruikt. Vulkanische rots van Gisenyi is daarvoor een veelbelovend adsorbens voor de verwijdering van zware metalen uit industrieel afvalwater.

Om het probleem van saturatie te omzeilen, werd een geïntegreerd systeem voor metaalverwijdering getest (hoofdstuk 6). Dit systeem is een combinatie van een *Upflow Anaerobic Packed Bed* (UAPB) reactor gevuld met vulkanische rots van Gysenyi welke tegelijkertijd dienst deed als adsorbens zowel als vasthechtingssubstraat voor bacteriën, en een vijver met waterhyacint. Het geïntegreerde systeem werd getest op de NUR campus (temperatuur 25 °C, maandelijkse instraling tussen 4,3 en 5,2 kWh/m^2). Er werden twee verschillende metaalvrachten gebruikt met respectievelijk lage (5 mg/L van elk metaal) en hoge (10 mg/L van elk metaal) metaalconcentraties in het afvalwater. Na een opstart- en adaptatieperiode van 44 dagen, werd elke metaalvracht getest gedurende 10 dagen bij een hydraulische verblijftijd van 1 dag. Goede verwijderingsefficiënties van minstens 86% werden genoteerd zowel bij lage als bij hoge metaalvrachten. De efficiëntie van de bioreactor werd niet wezenlijk beïnvloed door hogere metaalconcentraties in het influent. Een nageschakelde vijver met waterhyacint en onder een hydraulische verblijftijd van 1 dag, verwijderde een verdere 61% Cd, 59% Cu, 49% Pb en 42% Zn. Dit toont het belang als nazuiveringssysteem aan. Metalen accumuleerden voornamelijk in de wortels van de waterhyacint. De globale verwijderingsefficiëntie gemeten aan de uitlaat van het geïntegreerde systeem bedroeg 98% voor Cd, 99% voor Cu, 98% voor Pb en 84% voor Zn. Dit geïntegreerde systeem kan daarom gebruikt worden als een alternatieve zuiveringstechnologie voor metaalbevattende afvalwaters in ontwikkelingslanden.

De ontwikkeling van een goedkoop alternatief voor metaalverwijdering uit afvalwater heeft aangetoond hoe effectief een dergelijke technologie kan zijn. Metaalsulfide precipitatie als lange termijn verwijderingsmechanisme kan een efficiëntie van meer dan 90% bereiken in de bioreactor. Het geïntegreerde systeem heeft aangetoond hoe twee complementaire systemen voor metaalverwijdering in combinatie kunnen werken en een goede metaal verwijderingsefficiëntie kunnen bereiken.

Curriculum Vitae

Christian Sekomo Birame was born on 28 December 1968 in the city of Lubumbashi / Katanga province in the Democratic Republic of Congo (D.R.C). He did his primary school at Maadini school in D.R.C. He finished his high school in Kigali (Rwanda) at the Zairian Consular School in 1990. He joined the University of Lubumbashi in 1992 in the Institute of Applied Chemistry, where he did two years of undergraduate studies. After the 1994 genocide in Rwanda, he worked at the Ministry of labor and social affairs in 1995. In 1996, he joined the National University of Rwanda where he obtained a Bachelor of Science degree in Chemistry in 1998. From June 2000 to nowadays, he is working as a lecturer in the Chemistry department from the National University of Rwanda. From 2003 to 2005, he did his Master of Science in Electrochemistry at Witwatersrand University in Johannesburg (South Africa). From 2006, he started his PhD studies at UNESCO-IHE Institute for Water Education under a Capacity Building Project funded by the Netherlands Programme for Institutional Strengthening of Post-secondary Education and Training Capacity Project (NPT / RWA / 051-WREM). The research was carried out at the department of Chemistry and Civil engineering of the National University of Rwanda and at UNESCO-IHE (Delft, the Netherlands).

Publications and conferences

A. Publications

1. **Sekomo, C.B.**, Rousseau, D.P.L., & Lens, P.N.L., (2012a). Use of Gisenyi Volcanic Rock for Adsorptive Removal of Cd(II), Cu(II), Pb(II), and Zn(II) from Wastewater. *Water Air and Soil Pollution 223,* 533 - 547.

2. **Sekomo, C.B.**, Rousseau, D.P.L., Saleh, S.A.A.E., & Lens, P.N.L., (2012b). Heavy metal removal in duckweed and algae ponds as a polishing step for textile wastewater treatment. ***Ecological Engineering 44,*** 102 – 110.

3. **Sekomo, C.B.**, Pakshirajan, K., Rousseau, D.P.L, & Lens, P.N.L., (2012). Perspectives on textile wastewater treatment using physico-chemical and biological methods including constructed wetlands. *Submitted to the Journal of Chemical Technology and Biotechnology.*

4. **Sekomo, C.B.**, Nkuranga, E., Rousseau, D.P.L., & Lens, P.N.L., (2011a). Fate of Heavy Metals in an Urban Natural Wetland: The Nyabugogo Swamp (Rwanda). ***Water Air and Soil Pollution 214,*** 321 – 333.

5. **Sekomo, C.B.**, Kagisha, V., Rousseau, D.P.L., & Lens, P.N.L., (2011b). Heavy Metal Removal by Combining Anaerobic Upflow Packed Bed Reactors with Water Hyacinth Ponds. ***Environmental Technology.*** DOI:10.1080/09593330.2011.633564

6. Rousseau, D.P.L., **Sekomo, C.B.**, Saleh, S.A.A.E., & Lens, P.N.L., (2011) Duckweed and Algae Ponds as a Post-Treatment for Metal Removal from Textile Wastewater. In: Vymazal J. (Ed), ***Water and Nutrient Management in Natural and Constructed Wetlands***, Springer Science, Dordrecht. 63 – 75.

B. Conferences

1. **Sekomo, C.B.,** Rousseau, D.P.L., & Lens, P.N.L., (2010). Use of Gisenyi volcanic rock for adsorptive removal of Cd(II), Cu(II), Pb(II) and Zn(II) from wastewater. In: *Proceedings of the 2nd International Conference Research Frontiers in Chalcogen Cycle Science and Technology.* UNESCO-IHE / Delft, The Netherlands (25 – 26 May 2010).

2. **Sekomo, C.B.,** Rousseau, D.P.L., & Lens, P.N.L., (2009). Heavy Metal removal by a combined system of anaerobic Upflow Packed Bed Reactor and Water Hyacinth Pond. In: *Proceedings of the Ecological Engineering from Concepts to Application (EECA).* Cité internationale universitaire de Paris, France (02 – 04 December 2009).

3. **Sekomo, C.B.,** Rousseau, D.P.L., & Lens, P.N.L., (2009). Fate of Heavy Metals in an Urban Natural Wetland: The Nyabugogo Swamp (Rwanda). In: *Proceedings of the WETPOL 2009: 3rd Wetland Pollutant Dynamics and Control.* Caixa Forum and Hotel Barcelona Plaza, Barcelona, Spain (20 – 24 September 2009).

4. **Sekomo, C.B.,** Hutriadi, Saleh, S.A.A., Rousseau, D.P.L., & Lens, P.N.L., (2008). Heavy metals removal from a synthetic industrial wastewater. In: *Proceedings of the 1st International Conference G16 Research Frontiers in Chalcogen Cycle Science and Technology.* Wageningen, The Netherlands (28 – 29 May 2008).

T - #0112 - 071024 - C150 - 244/170/8 - PB - 9780415641586 - Gloss Lamination